CONSUMER PRODUCT INNOVATION AND SUSTAINABLE DESIGN

The evolution and impacts of successful products

Robin Roy

LONDON AND NEW YORK

First published 2016
by Routledge
2 Park Square, Milton Park, Abingdon, Oxon OX14 4RN

and by Routledge
711 Third Avenue, New York, NY 10017

Routledge is an imprint of the Taylor & Francis Group, an informa business

© 2016 Robin Roy

The right of Robin Roy to be identified as author of this work has been asserted by him in accordance with sections 77 and 78 of the Copyright, Designs and Patents Act 1988.

All rights reserved. No part of this book may be reprinted or reproduced or utilised in any form or by any electronic, mechanical, or other means, now known or hereafter invented, including photocopying and recording, or in any information storage or retrieval system, without permission in writing from the publishers.

Trademark notice: Product or corporate names may be trademarks or registered trademarks, and are used only for identification and explanation without intent to infringe.

British Library Cataloguing-in-Publication Data
A catalogue record for this book is available from the British Library

Library of Congress Cataloging-in-Publication Data
Roy, Robin.
Consumer product innovation and sustainable design : the evolution and impacts of successful products / Robin Roy.
pages cm
Includes bibliographical references and index.
1. Product design. 2. Sustainable design. 3. New products. 4. Technological innovations. I. Title.
TS171.R687 2016
658.5'75--dc23
2015016983

ISBN: 978-0-415-86997-3 (hbk)
ISBN: 978-0-415-86998-0 (pbk)
ISBN: 978-1-315-71972-6 (ebk)

Typeset in Bembo
by Fakenham Prepress Solutions, Fakenham, Norfolk NR21 8NN
Printed by Ashford Colour Press Ltd, Gosport, Hants

**658.
575
ROY**

CONSUMER PRODUCT INNOVATION AND SUSTAINABLE DESIGN

Consumer Product Innovation and Sustainable Design follows the innovation and evolution of consumer products from vacuum cleaners to mobile phones from their original inventions to the present day. It discusses how environmental concerns and legislation have influenced their design and the profound effects these products have had on society and culture. The book also uses the lessons from the successes and failures of examples of these consumer products to draw out practical guidelines for designers, engineers, marketers and managers on how to become more effective at product development, innovation and designing for environmental sustainability.

Robin Roy is Emeritus Professor of Design and Environment at the Open University. Since joining the OU in 1971 as one of the first lecturers in Design, he has chaired and contributed to many OU courses on design, innovation, energy and environment, most recently *Design Essentials; Innovation: Designing for change*; and *Environment: Journeys through a changing world*. In 1979, he founded the Design Innovation Group to research design and innovation management and sustainable design. He has published many books, book chapters, papers and articles on topics ranging from design creativity and the successful management of new product development to environmentally sustainable education systems and consumer adoption of low and zero carbon technologies. He is a Fellow and Council member of the Design Research Society, a former Director of Carbon Connections Ltd. and a Trustee of Powerful Information, a local international development charity.

'Robin Roy has written intriguing histories of a selection of consumer products – washing machines, lamps and lighting, televisions, vacuum cleaners and mobile phones – looking at the patterns of their technical evolution and the causes of success and failure. He puts special emphasis on the environmental impacts of designs including their use of energy and materials. Professor Roy with his lifetime experience of studying product innovation is uniquely qualified to write this book that will fascinate the general reader and be particularly useful in teaching, design and industry.' – *Professor Philip Steadman, Energy Institute, University College London, UK*

'*Consumer Product Innovation and Sustainable Design* is a fascinating and incredibly useful book that builds upon a significant amount of research from the author. Its focus around five types of products provides the reader with both a set of interesting histories but also design case studies which will be particularly useful for students and practitioners. The focus on sustainable design within these case studies helps the reader understand the subject within the context of product innovation. This book should become essential reading for all product designers no matter what career stage they are at.' – *Professor Tracy Bhamra, Loughborough University, UK*

'By combining the archives of *Which?* Magazine with an analysis of the evolution of popular consumer product categories *Consumer Product Innovation and Sustainable Design* reminds us why the things around us have the capacity not just to shape our own lives, but that increasingly it is how a product uses our limited resources efficiently that is the key to success. Robin Roy has produced a book that will be invaluable to product designers and others in thinking innovatively and responsibly in determining tomorrow's good designs.' – *Peter Lloyd, Professor of Design and Associate Dean, University of Brighton, UK*

'*Consumer Product Innovation and Sustainable Design* draws on a huge amount of research and experience from the author and is based on a series of solid cases on the innovation process associated with four consumer products. Of particular note is integration of environmental considerations into the analyses and the final section on lessons for product developers and innovators. This book will be an essential resource for both practitioners and researchers interested in learning lessons on the evolution of innovation of 'active' consumer products.' – *Professor Martin Charter, Director, The Centre for Sustainable Design, University for the Creative Arts, UCA Farnham, UK*

'Robin has taught and researched consumer product innovation and sustainable design for many years. We have to thank him for producing this timely book that gives all of us access to a great archive of information set within a framework that challenges contemporary design practice.' – *Professor Tom Inns, Director, Glasgow School of Art, UK*

CONTENTS

List of illustrations ix
Preface xv
Acknowledgements xix

CHAPTER 1
INTRODUCTION: PATTERNS OF INNOVATION 1
Why this book? 2
How *Which?* tests consumer products 2
The consumer product case studies 4
Innovation and evolution of the bicycle 4
Theories of technological and design change 14
Practical lessons for product developers 17
References 17

CHAPTER 2
WASHING MACHINES 21
Washing machine technology and design 22
Environmental impacts and regulation 29
Laundry innovation 35
Social influences and impacts 37
Future developments 39
The pattern of innovation 44
References 44

CHAPTER 3
LAMPS AND LIGHTING 47
Electric lamp technology 47
Lamp and lighting design 59
Environmental impacts and regulation 60
Social influences and impacts 64
Future developments 65
The pattern of innovation 68
References 68

CHAPTER 4
TELEVISION 71
Television technology 71
Television design 83
Environmental impacts and regulation 87
Social influences and impacts 90
Future developments 92
The pattern of innovation 93
References 94

CHAPTER 5
VACUUM CLEANERS 97
Vacuum cleaner technology and design 98
Environmental impacts and regulation 114
Social influences and impacts 115
Future developments 116
The pattern of innovation 117
References 117

CHAPTER 6
MOBILE PHONES 119
Mobile phone technology 119
Operating systems 124
Mobile phone design 124
Environmental impacts and regulation 132
Ethical and political issues 136
Social influences and impacts 137
Future developments 140
The pattern of innovation 141
References 142

CHAPTER 7
LESSONS FOR PRODUCT DESIGNERS, DEVELOPERS AND INNOVATORS   147

Understand patterns of innovation	147
Follow new scientific knowledge and enabling technologies	156
Designing for product success	157
Avoiding product failure	165
Balance technology push and market pull	168
Designing for the environment	169
Understand social influences and impacts	175
Designing for the future	176
References	178
Illustration acknowledgements	181
Index	185

LIST OF ILLUSTRATIONS

Figures

1.1 *Which?* Best Buy and other product assessment logos
1.2 Ordinary, high-wheel or penny-farthing bicycle, 1870s
1.3 Third prototype Rover safety bicycle with diamond frame and solid tyres, 1885
1.4 Classic diamond-frame sports bicycle with an alloy steel frame, multiple derailleur gears and 27-inch wheels with aluminium rims, 1990s
1.5(a) Alex Moulton with three models of his small-wheel bicycle with rubber suspension, 1967
   (b) Alex Moulton riding his 'advanced engineering' space frame bicycle, 1983
1.6 Brompton folding bicycle
1.7 Streamlined human-powered vehicle (HPV)
1.8 Mike Burrows' Mark 1 racing bicycle with aerodynamic monocoque (single shell) carbon fibre frame, 1985
1.9 Hybrid bicycle, a general purpose or commuting bicycle that combines features of mountain, road and touring bikes, *c.* 2005
1.10 Bianchi Sempre road and racing bicycle with aerodynamic carbon fibre frame, 2012
1.11 Smart electrically-assisted bicycle
2.1(a) Jacob Schaffer's washing machine, Germany, 1766
   (b) Edward Beetham's washing mill, England, 1790s
2.2(a) Improved Doty hand-operated squeezer washing machine, New York, *c.* 1870
   (b) 'Pioneer Masher' manual washing machine and mangle, Accrington, England, 1903

2.3   Thor washing machine, Chicago, c. 1907, designed by Alva J. Fisher and patented 1910, was one of the first electric washing machines. The exposed electric motor under the tub sometimes caused hazardous electric shocks
2.4   Electric washing machine with a powered wringer, made by Beatty Bros of Canada, c. 1920. The wooden tub with a 4-legged rotating wooden dolly inside is like those of earlier manual machines. The exposed motor could be hazardous
2.5   Maytag Model 18 top-loading washing machine with wringer, USA, 1939
2.6   The first mass-produced, fully automatic washing machine, the Bendix Home Laundry was first launched in 1937
  (a) Scott's automatic washing machine patent
  (b) Bendix automatic, c.1945
2.7   Modern Indian twin tub washing machine, 2015
2.8   Some automatic washing machines on UK market in 1966
  (a) Bendix
  (b) English Electric horizontal axis front loading
  (c) Hoover Keymatic inclined axis front loading
  (d) Hotpoint vertical axis top loading
2.9   Some *Which?* Best Buy automatic washing machines on UK market in 1994. Left to right: Candy Aquaviva; Zanussi washer drier; Zanussi Jetsystem
2.10 Percentage contributions of the production, distribution, use and disposal phases of a washing machine's life cycle to its environmental impacts, excluding detergent manufacture and distribution. (Adapted from Roy (2006) p. 132 and Durrant *et al.* (1991) Figure 2.3, p. 37)
2.11(a) Industrial design sketch of the Hoover New Wave washing machine controls
  (b) The New Wave range was awarded the first EU Ecolabel
2.12(a) The EU washing machine energy label (2010 on) provides energy and spinning efficiency ratings, annual electricity and water use and noise output
  (b) Bosch washing machine with an A+++ energy label, 2014
2.13(a) Samsung Ecobubble™ washing machine, 2013
  (b) Dyson Contrarotator™ washing machine with two drums that rotated in opposite directions to simulate hand-washing, 2000–2005
2.14 Xeros large commercial washer that uses small amounts of water and detergent to lift dirt from the wash, which is then absorbed and removed by reusable polymer beads
2.15 Concept design for a sustainable laundry service (From a draft version of Dewberry *et al.*, 2013, Figure 3)
3.1   Pre-electric lamps
  (a) nineteenth-century oil lamp

	(b) Ancient to nineteenth-century oil and gas lamps
3.2	Thomas Edison's patent drawing and application for an improvement in electric lamps, patented 27 January 1880
3.3(a)	Early twentieth-century tungsten filament incandescent light bulbs
(b)	Early twenty-first-century clear and frosted tungsten filament incandescent (GLS) light bulbs with bayonet and screw fittings yes, 'Modern incandescent light bulbs'
3.4	Modern tungsten halogen incandescent light bulbs (left) GLS light bulb with tungsten halogen capsule (right) Tungsten halogen reflector spot lamp with GU10 connector
3.5	Circular fluorescent lamps – decorative, but much less common than straight fluorescent tubes
3.6	Compact fluorescent lamps (a) Philips SL18, the first mass-produced CFL, 1980 (b) Modern CFL lamps with screw and bayonet connectors and CFL spot lamp with GU10 connector
3.7	LED lamps with standard screw and bayonet connectors and 16-LED spot lamp (a direct replacement for a GU10 halogen spot lamp)
3.8(a)	Stick CFL desk lamp
(b)	LED wardrobe light that comes on automatically when the door is opened
3.9	Life cycle energy use of GLS and halogen incandescent, CFL and LED lamps (Adapted from: US DOE, 2013a)
3.10	EU Energy Labels for halogen incandescent (C-class), CFL (A-class) and LED (A+class) light bulbs
3.11	Philips DNA Helix concept chandelier using organic light emitting diode (OLED) panels, 2012
4.1	Baird Televisor electro-mechanical television receiver with a rotating disc scanner and 3-inch screen, 1930
4.2	Baird T-23 15-inch cathode ray tube mirror lid TV, 1938
4.3(a)	RCA CT-100 15-inch colour television, one of the first American receivers with a colour CRT, 1954
(b)	Sony 18-inch colour TV with Trinitron cathode ray tube, c. 1970
4.4(a)	JCV Video Home System (VHS) video-recorder, 1976
(b)	Sony Betamax video-recorder, 1979
4.5	Sky+ personal digital video recorder (PVR). It can record two standard or high-definition programmes on its 500GB to 2TB hard disk while showing a third programme, as well as providing access to TV catch-up and Internet TV
4.6	RCA TRK-9, a TV receiver with a 9-inch diagonal cathode ray tube and radio housed in 'streamlined' veneered wooden cabinet, 1939
4.7(a)	Baird 12-inch 405-line black and white CRT TV, 1949. It was available as the Townsman, for reception close to the transmitter, and the Countryman for fringe areas

(b) Bush TV12 with 9-inch 405-line black and white CRT and Bakelite case, 1948
(c) GEC TV set with 14-inch 405-line black and white CRT, 1955
4.8 Sony 'big box' 32-inch widescreen cathode ray tube colour TV, c. 2005
4.9(a) Panasonic 32-inch screen flat panel liquid crystal display (LCD) TV, 2005
(b) Samsung 40-inch screen light emitting diode (LED) Smart TV, 2013. Note the narrow screen border compared to Figure 4.9a
4.10(a) Samsung 22-inch (54 cm) screen LED Smart High Definition TV, 2014
(b) The Samsung 22-inch TV's 'A' energy efficiency rating, power and annual energy use on its EU Energy Label
4.11 Samsung 55-inch '4K' Ultra-High Definition curved screen LED smart TV
5.1 Bissell carpet sweeper late 1870s (middle left) and various designs of late nineteenth and early twentieth-century bellows or bicycle pump-operated manual vacuum cleaners
5.2 Booth's 'Puffing Billy' horse-drawn vacuum cleaner, 1901. The machine was parked outside the house to be vacuumed and the cleaning hoses fed through the windows. It was originally powered by an oil engine inside the carriage; an electric motor was substituted later
5.3(a) Spanger's vacuum cleaner patent, 1908
(b) Hoover (Electric Suction Sweeper Company) Model 0, 1908
5.4 (left) British Magic Appliances vacuum cleaner; (middle and right) Hoover upright cleaners, 1919 model and Hoover Junior 1936
5.5 Electrolux Model V cylinder vacuum cleaner, 1921
5.6 Electrolux Model XXX (30) cylinder cleaner, 1937, styled by American industrial designer, Lurelle Guild
5.7(a) Advertisement for Hoover Model 29 in bright red, 1950, 'styled by Henry Dreyfuss'
(b) Hoover Constellation 1958, which floated on a cushion of its own exhaust air and remained in production until 1980. The distinctive spherical design again by Henry Dreyfuss reflected late 1950s American concerns with space following the launch of the Russian Sputnik 1 satellite in 1957
5.8 Typical late 1970s vacuum cleaners on the British market:
(a) Hoover Dirtsearcher Junior U1016
(b) Hoover Convertible U5010
(c) Electrolux Automatic 345
(d) Hoover Celebrity S3005, an updated version of the floating Constellation (Figure 5.7b)
5.9 Recent model of the hand-held, rechargeable battery-powered Black and Decker Dustbuster, first launched in the USA in 1979
5.10(a) (left) Dyson's first production cleaner with concentric cyclones, the G-Force, made in Japan; (centre) the first prototype in which the

cyclones were placed side by side; (right) a few of the models used by Dyson to develop the best shape of cyclone

(b) Dyson's pink and lavender G-Force dual cyclone vacuum cleaner, 1986, made and sold in Japan

5.11(a) Dyson DC01 Dual Cyclone bagless upright cleaner, made and launched in the UK in 1993

(b) Dyson DC02 bagless cyclonic cylinder cleaner, 1995

5.12 Dyson DC24 multi-floor lightweight upright cleaner with several root cyclones and highly manoeuvrable Dyson Ball™ steering, 2012

5.13(a) Form follows function: Dyson DC75 Cinetic Big Ball with multiple oscillating tip cyclones to remove the finest particles without a filter, 2015

(b) Function plus sleek styling: Miele Dynamic U1 Powerline upright bagged vacuum cleaner, with a 1500W motor that meets the requirements of the EU Energy Label, 2014

5.14 Dyson 360 Eye™ robot vacuum cleaner, 2015

6.1 Most telephone communications in developing and newly industrialised countries, such as India, are by mobile phone

6.2 Motorola 4500x 'transportable' car phone, 1988

6.3 The first commercially available hand-held portable mobile phone, the Motorola DynaTAC 8000X, 1983. This is a 1984 model with a red LED display

6.4 The first mass-produced 'candy bar' mobile phone was the Nokia 1011 GSM digital phone, 1992. This is a 1993 model

6.5 The first clamshell phone, the Motorola StarTAC, 1996

6.6 Desirable Nokia 8110i GSM slide phone, 1997, which featured in the film *The Matrix*

6.7(a) Nokia 3310 GSM candy bar phone, of 2000 was an iconic design that sold 126 million units worldwide

(b) Nokia 1100 basic GSM candy bar phone 2003, designed with a dustproof case for developing countries, sold about 250 million units worldwide

6.8 The highly fashionable Motorola RAZR V3 ultra-slim clamshell phone, 2004

6.9 The first touchscreen mobile phone was the 2G Apple iPhone, launched in 2007 with a 3.5-inch screen. (left) a 3G version of the original iPhone, 2008 (right) a 5.5 inch screen iPhone 6 Plus, 2014

6.10 Dominant touchscreen smartphone designs
(a) 5.5 in iPhone 6 Plus, 2014
(b) rivalled by one of the Samsung Galaxy range, the S5, 2014

6.11 Blackberry Q10 smartphone with physical QWERTY keyboard

6.12 Nokia 100 budget phone, 2013

6.13 Mobile handset design forms appearing in a sample of the magazine T3 (*Tomorrow's Technology Today*) between 1996 and 2009 showing the divergence and convergence of designs (From Muir Wood, 2010, p. 132)

6.14   Design award winning Sony Ericsson T610 camera-phone, 2003
6.15   Spoof advertisement from the Gaia Foundation highlighting the negative impacts involved in the extraction of materials for making mobile phones. The 'apps' on this phone are: resource depletion, ecosystem destruction, land grabbing, inbuilt obsolescence, toxic waste, conflict minerals, poor working conditions
6.16   Google Glass with an optical head-mounted display, providing many computer and communications functions in a wearable product, especially when paired with a smartphone. It was available from 2013 to 2015 when the 'Project Glass' experimental phase ended and new versions were under development
7.1   Typical pattern of product and process innovation over time. (Adapted from Utterback and Abernathy (1975), p. 645 and Abernathy (1978), p. 72)
7.2   Technological S-Curve: diminishing improvements in performance over time result in competition between, and eventual displacement of, existing dominant designs by innovative products based on a disruptive technology (Adapted from Utterback (1994), p. 160.)
7.3   Philips LaserDisc player (1982) failed to catch on because of its higher price and perceived lack of relative advantage over videocassette recorders (VCRs)
7.4   Different approaches to, or levels of, designing for the environment. (Adapted from Brezet (1997), p. 22 and Roy (2006), p. 90)

Tables

3.1   Advantages and disadvantages of different energy-efficient lamps
3.2   Efficiency, life and life-time cost of typical electric lamps

PREFACE

I have always been interested in the design of things, but as a child I was especially fascinated by the historical technologies and products in the Children's and Domestic Appliances galleries at the Science Museum in London. Later this led me to study undergraduate mechanical engineering and then take a master's degree and PhD in design technology at Manchester University under Christopher Jones, one of the originators of systematic design methods and a broad systems approach to design. After briefly working as a trainee in engineering companies in Britain and Sweden, I was fortunate to be appointed in 1971 as one of the first lecturers in Design at the newly founded Open University (OU). The OU pioneered teaching large numbers of home-based adult students via specially produced books, radio and television programmes, supported by local tutorials and residential schools. The Design group at the OU developed an interdisciplinary and project-based approach to teaching design at a distance, which it has maintained to the present day, but now using online multimedia as well as printed books. For my research I founded the Design Innovation Group in 1979, which focused on investigating the successful practice and management of product design and technical innovation in industry and on developing the new field of sustainable design. This research often arose from the work involved in developing the OU distance learning courses and making the associated television programmes, audios and videos.

As part of my interest in products and their evolving design, I subscribed to the British consumer magazine, *Which?* This publication provides a unique record of how consumer products available in Britain have changed over almost 60 years. In my position as an Emeritus Professor of Design and Environment at the OU, I had the opportunity to make use of my *Which?* archive. The idea for this book was to use my *Which?* collection as a way of tracking the innovation and evolution of a number of consumer durable products, such as washing machines, vacuum cleaners, television equipment and mobile phones. However, I quickly realised that although *Which?* was a useful basic source, tracking the evolution of such products from their invention in the nineteenth and early twentieth centuries to the present required a considerable amount of additional research and scholarship making use of many other printed and online resources and visiting museums and shops. Arising from my work on sustainable design I was also interested in discovering whether, why and when environmental concerns and regulations influenced the design of these products; what socio-economic and cultural factors influenced their development; and what the products' impacts on the environment and society were.

My aim in writing the book was to use the product case studies to provide general conclusions about patterns of technical innovation and design evolution and the extent to which these patterns fitted existing theories of innovation and design. A more practical aim, linked to my earlier work on successful design and innovation, was to make use of the empirical information from the product case studies to provide lessons and guidelines for designers, engineers, innovators, managers and marketers, as well as for educators of these professions. I hope that these groups will find the book useful as well as interesting.

In writing the book I have discovered that there are many individuals who are so fascinated with certain technologies and products that they collect them, make websites cataloguing and describing them, and even set up museums to show them to the public. There are collections, websites and museums for washing machines, vacuum cleaners, lamps, televisions and mobile phones, which I have found to be most valuable in writing this book. I hope that such collectors, and people who are simply curious about the history and design of consumer products, might find this book interesting and useful too.

Robin Roy
Milton Keynes, February 2015

For June, Alex, Linda, Ella, Joe and Olivia

ACKNOWLEDGEMENTS

I wish to thank Linda Wolfe and Pam Matthews of Intertek, Milton Keynes (the former Consumers' Association test laboratory) for giving me access to Intertek's archive of *Which?* magazines to fill gaps in my collection. I am also most grateful to Intertek's product specialists, Paul Rogers; Jeremy Owens; Ian Mann; Michael Meed and Stephen Higgins, and also Tim Lister of IBR UK Ltd., who commented on drafts of the case study chapters. Barry Dagger, former editor of *Engineering Designer*, also provided useful advice and comments.

Many thanks are also due to Carol Houghton, who provided expert assistance with picture research and rights clearances, and to Andrew Metcalf for graphic design work. Lee Maxwell provided valuable photographs from his vast collection and museum of washing machines and Steve McVoy kindly provided pictures of historic televisions from the Early Television Museum. I am grateful to my editors at Routledge: Kat Holloway, who commissioned the book; Emma Gadsden and Grace Harrison, who kept me going; and Trudy Varcianna and Hannah Champney, who saw the book through to completion. Thanks also to Nicola King, a researcher for Intertek who used to work for *Which?*, for expertly compiling the index.

I want to especially thank my partner, June Payne, for proofreading and commenting on draft chapters and for patiently waiting for jobs in the house and garden to be done while I spent many months researching and writing the book.

1
PATTERNS OF INNOVATION

The inspiration for this book was my collection of the British consumer magazine, *Which?* The magazine publishes independent test reports, advice and other information to the over 640,000 consumers who currently subscribe to it to help them choose goods and services that perform well and offer good value for money. *Which?* magazine publishes test reports and articles on goods and services ranging from consumer durables such as washing machines, televisions and light bulbs and consumable items such as food, batteries and laundry detergents to services such as hospitals, banks, airlines and financial advisers. Since 1996 the scope of the magazine has been supplemented and extended by reports and advice on a similar range of goods and services on the *Which?* website.

The magazine was first launched in London in 1957 by the then newly founded Consumers' Association and has been published continuously ever since. It is now published by Which? Ltd., the commercial arm of the Which? Group. The Consumers' Association, the charitable arm of the Group, carries out other activities, such as carrying out the product tests and campaigning on consumer issues.

For my research and teaching in the field of product design, I have for many years studied the technical and design evolution of consumer products such as washing machines and bicycles (see e.g. Roy, 1980; Roy, 1999; Roy and Tovey, 2012). I therefore kept my collection of *Which?* magazines because it, together with the *Which?* website, provides a unique written and pictorial record of how such products have changed over almost 60 years, which I felt would provide the basis for an interesting and useful book. As well as identifying patterns of innovation and evolution of consumer products, I was interested in whether, when, why and how environmental criteria such as improved energy efficiency became part of their technical specification. I also wanted to discover more about the socio-economic, cultural and political influences on and impacts of consumer product innovations (see e.g. Roy, 1994; Smith, Roy and Potter, 1996; Roy, 1997).

2 Patterns of innovation

Why this book?

My idea for the core of the book was a series of case studies that would track the innovation and evolution of different classes of consumer product, including the introduction of ecodesign features and sustainability criteria into their specification and their impacts on the environment and society.

The case studies would be based on *Which?* magazine and online as major sources of information, together with other primary and secondary print and online sources and information obtained from visits to shops and museums and discussions with retailers.

Another aim of the book was to use empirical information drawn from the product case studies to provide general conclusions about patterns of technical innovation and design evolution and the extent to which this information supported existing theories of innovation and design. However, a particular intention of the book was to make use of the evidence of the case studies to provide useful information and lessons for product designers, engineers, developers, managers and marketers and for educators of these professions. These lessons would, for example, provide guidelines on what makes some models and brands of consumer product successful and others market failures; how to design products for reduced ecological impacts; how to take consumer preferences and social behaviour into account in product design, and any general trends that might be useful when planning future consumer products.

How *Which?* tests consumer products

Before choosing and developing the product case studies it was important to understand how the reports published in *Which?* were produced. This was done by looking at issues of the magazine from 1957 to 2014, looking at the *Which?* website and consulting Consumers' Association annual reports. How the Consumers' Association conducts its product tests and *Which?* produces its reports is outlined in the box below.

Which? evaluation methods

To produce its test reports the Consumers' Association first buys the products to be tested from ordinary retailers and then employs a variety of methods to evaluate them. The methods have evolved over the years, but from the early days they have included laboratory tests to provide objective measures of technical performance; for example, how well a washing machine cleans standard samples of fabric stained with various substances (e.g. oil, ink, cosmetics). Another long-established method is obtaining the views, originally of Consumers' Association staff then of panels of experts, on products in use in the laboratory; for example, their opinions on the picture and sound quality of a TV set. Other evaluation methods include trials of products under controlled conditions by consumers chosen from panels of Consumers' Association members. Such trials might involve, for example, users with young children steering different push-chairs around a standard obstacle course in the test lab and/or using the products at home or in other normal environments and then completing a questionnaire on their experience. The Consumers' Association has also for many years conducted surveys of its members, for example, on product reliability, and gathered consumer feedback on specific products via questionnaires or online.

The results of these various evaluation methods, plus information on prices, product specification and features, are then analysed. Conclusions on which products consumers are recommended to buy are provided in *Which?* magazine and more recently also online. Over time a system has evolved to combine the various test results and evaluations, which started with a blob or star rating for different measures of performance, convenience, etc. Later a rating that gives a weighted overall score for important consumer choice factors was introduced. For example, a 1995 report on washing machines weighted the test results for cleaning performance, running cost, ease of use, creasing and spinning efficiency to give an overall score out of 10 (Consumers' Association, 1995). This system then evolved into one that gives an overall percentage score for each product tested. For example, the score for a washing machine is currently based on combining its percentage scores for cleaning performance (up to a maximum of 50%); rinsing (max. 15%); spin drying (max. 15%); energy use (max. 10%); programme time (max. 5%) and ease of use (max. 5%).

From its early days *Which?* has recommended products that perform well in its tests, naming it either as a 'Best Buy' or as offering 'Good Value for Money'. Today, products that score especially well in the comparative tests and trials – and for which there is no evidence of poor reliability – are still recommended as a Best Buy. These account for approximately the top 20% of products in terms of their performance within a category. Some energy-using products, such as refrigerators, dishwashers and TV sets, that combine a Best Buy performance with high energy efficiency measured in its laboratory tests, are given an 'Energy Saver' logo. Other symbols that appear in *Which?* magazine reports include 'Worth a Look' for products that, while not necessarily a Best Buy, performed well in tests, 'Great Value' for products that in addition offer excellent value for money, and 'Don't Buy' for especially poor performing, unhealthy or unsafe products (Figure 1.1).

FIGURE 1.1 *Which?* Best Buy and other product assessment logos

The consumer product case studies

Given the very wide range of consumer durable products that Which? tests and reports on, one of my first tasks was to choose which ones to focus on for this book. I wanted products that represented different levels of technological complexity, different rates of technological and design change, differences in the importance of engineering, aesthetics and ergonomics in their design, and examples of mechanical, electro-mechanical and electrical/electronic products. I also wanted to uncover different social, cultural and behavioural influences on design and innovation and different levels of impact of the products on the environment and society.

My long list of possible product classes included: white goods (washing machines, dishwashers, refrigerators and freezers); brown goods (radios, television and audio equipment); consumer electronics (personal computers, mobile phones and cameras); other domestic appliances (cookers, vacuum cleaners, coffee makers); personal transport (bicycles, prams and push-chairs) and lighting (lamps and light fittings).

The above criteria were then used to select a short list of product classes for the case studies that would form the core content of the book. My final choice also took into account how much the product had been reported on in *Which?*; the amount of new research that would be involved in developing the case, and likely duplication in the lessons that may be drawn from it and other cases.

The final set of product case studies I chose is: washing machines (Chapter 2); lamps and lighting (Chapter 3); television equipment (Chapter 4); vacuum cleaners (Chapter 5) and mobile phones (Chapter 6). Chapter 7 then presents the general conclusions and the practical lessons for product developers, managers and educators based on the previous chapters.

Innovation and evolution of the bicycle

In order to introduce the patterns of change and types of lessons I want to cover and to provide a framework for researching the other case studies, I will start with a case study of the invention, innovation and evolution of the bicycle; a consumer product with a very long history, which I have studied before (see e.g. Roy and Tovey, 2012).

CASE STUDY: THE BICYCLE

The invention, development and evolution of the modern bicycle can be viewed as consisting of three phases: a divergent experimental phase, a dominant design phase and an innovative design phase.

The divergent experimental phase

The origin of the bicycle is unknown. What is known is that in 1818 Karl Freiherr Drais von Sauerbronn, a German baron and professor of mechanics, was granted a patent on a running machine, which consisted of a wooden frame to which two iron-tyred, wooden wheels were attached. The rider propelled the so-called 'Draisienne' by taking long strides along the ground and steered by means of a handlebar fixed to the front fork and wheel.

The Velocipede

The first 'proper' bicycle with pedals, the Velocipede, was invented in France by Pierre and Ernest Michaux and introduced in 1861. It consisted of a light-weight Draisienne propelled via pedals and cranks fixed to an enlarged front wheel. It is said that the Michaux got their idea for pedals by thinking of an *analogy*; the handles on a grinding wheel. The Velocipede was a key invention and innovation, but bicycles of this type became known as 'bone-shakers' because their rigid cart-type wheels gave a rough ride and their weight and direct drive limited speed.

High-wheel bicycles

The next major step in bicycle evolution occurred in 1870 when James Starley and William Hillman patented the Ariel bicycle. The Ariel's key innovation was its lighter, less rigid wheels with wire spokes and rubber tyres that gave it a smoother ride. Also, the Ariel's enlarged front wheel allowed the rider to drive the bicycle much faster than the Velocipede.

So-called Ordinary or high-wheel bicycles then took the large front wheel design to its logical conclusion. This wheel reached the greatest possible diameter for direct pedal drive dictated by the rider's leg length, while the rear wheel shrank in size, hence the nickname 'penny-farthing' (Figure 1.2). Its advantages of speed and simple design made the Ordinary bicycle very popular, mainly among young men willing to accept the risk of injury or death due to its tendency to throw the rider over the handlebars if the front wheel hit an obstacle. However, difficulty of mounting and danger of riding a high-wheel bicycle stimulated numerous attempts to design safer and easier-to-use cycles for other potential users such as older men and women.

Safety cycles

'Safety' cycles were of three types: Ordinary bicycles with an indirect chain or treadle drive to smaller front wheels; various designs of tricycle and multi-wheel cycles; and bicycles driven via pedals and chain to the rear wheel – like a modern bicycle.

Archibald Sharp (1896) provided a comprehensive catalogue of nineteenth-century cycle types classified according to the number of wheels, the method of steering and the means of transmission. He identified 24 types of bicycle, 21 types of tricycle, and several types of monocycle, dicycle (wheels side-by-side) and multicycle (four or more wheels) that were manufactured from 1865 to 1890. During this divergent experimental phase numerous technically unsound and even crazy designs – such as a huge single-wheel machine with the rider sitting inside it – were created by inventors. Even the first rear chain-driven bicycles had unnecessarily large rear wheels because their designers did not understand the concept of gearing up via the chain. This led Sharp to lament the lack of engineering knowledge of many early cycle designers which had led to a plethora of poor designs. Early chain driven bicycles were also mocked for their ungainly appearance by adherents of the more elegant high-wheel designs.

The Rover Safety

A key step in the evolution of the modern bicycle was the 'Rover' Safety bicycle designed in 1884 by John Starley. Like most innovations the Rover did not emerge 'out of the blue', but was the result of nearly 20 years of cycle and component development. The developments included wire-spoke wheels, ball bearings, roller chains and numerous other aspects of earlier designs. Starley built the first prototype of the Rover Safety in 1884. It had indirect steering and a 3-foot (approx. 1 m) front wheel as Starley seemed unable to completely break away from the high-wheel form. It was not until the third prototype of 1885 (Figure 1.3) that a diamond-shaped frame and direct steering to the front forks appeared. These were ideas first applied in Humber's tricycle and safety bicycle of 1884. Like most creative individuals, Starley put together what had gone before in a better way. However, what led him to conceive the form of the Rover Safety was an *analogy*. Starley likened someone exerting the maximum force on the pedals to them climbing a ladder and realised that placing the handlebars, saddle and pedals in the correct configuration to allow this required a rear chain driven design (Starley, 1898).

FIGURE 1.2   Ordinary, high-wheel or penny-farthing bicycle, 1870s

FIGURE 1.3   Third prototype Rover safety bicycle with diamond frame and solid tyres, 1885

The pneumatic tyre

Dunlop's pneumatic tyre was a highly significant innovation applied to the Safety bicycle. The pneumatic tyre not only increased the rider's comfort by absorbing road bumps, but also allowed increased speeds by reducing the energy lost through vehicle vibration. Dunlop had been considering the problem of vehicle vibration for many years. Eventually his experience as a veteran of making rubber surgical appliances gave him the idea of making an air-filled rubber tyre – Dunlop's creative thinking thus involved the *transfer* and *adaptation* of an existing technology (Roy and Tovey, 2012). Dunlop tested his idea by fitting wheels with prototype air tyres to his son's tricycle.

This showed that pneumatic tyres worked well, so Dunlop fitted a second tricycle with improved tyres, which provided an especially smooth and fast ride. In 1888 he therefore applied for a patent on his invention. As with early chain-driven bicycles, pneumatic tyres were considered ugly by purist cyclists, but when bicycles with pneumatic tyres won most races, they had to admit their superiority over solid rubber and cushion tyres.

The classic diamond-frame bicycle

By about 1890 the bicycle had converged onto its classic diamond frame, rear chain-driven form, similar to a modern bicycle. This classic design, when fitted with pneumatic tyres, became a very efficient and comfortable machine and led to a bicycling boom, which in the mid-1890s spread through the middle and upper classes, and to both men and women.

The dominant design phase

The classic diamond-frame bicycle fulfilled its function so well that, although cycle inventions did not cease after the mid-1890s, few novel designs entered the market. The diamond-frame bicycle thus became the *dominant design*. The attention of manufacturers therefore turned to innovations in *production* technology to reduce costs and to increase output to meet the demand created by the bicycling boom. After the boom was over, major firms like Raleigh only survived by adopting further production innovations to reduce costs.

During the twentieth century, the dominant diamond-frame bicycle design underwent a process of incremental innovation and improvement. Bicycles

FIGURE 1.4 Classic diamond-frame sports bicycle with an alloy steel frame, multiple derailleur gears and 27-inch wheels with aluminium rims, 1990s

evolved into a range of modern products for transport, leisure, touring, racing, etc. The evolution of the dominant design involved the adoption of stronger and lighter frame materials and new and improved components including gears and cable brakes (Figure 1.4).

During the first half of the twentieth century the bicycle became a consumer product used mainly for sport and leisure and, when car ownership increased, by those who couldn't afford any other form of transport. However, a turning point came in the 1960s when Alex Moulton launched his small-wheel bicycle, which arguably helped to trigger a second divergent phase of cycle design.

The innovative design phase

Moulton small wheel bicycles

Moulton's interest in bicycles was revived by the petrol shortages resulting from the 1956 Suez crisis. Moulton questioned whether the 27-inch (70 cm) wheels of the dominant bicycle design were necessary, as he considered that vehicles with small wheels, like the Mini car which he helped to design, were in the direction of design evolution. Moulton therefore began developing a bicycle with 16-inch (40 cm) diameter wheels, which was launched in Britain in 1962. Moulton considered that a suspension system was essential for his bicycle because small wheels are less able to absorb road shocks and fall deeper into holes than a large wheel. Given his background in designing rubber vehicle suspensions for the Mini car, it is not surprising that Moulton *transferred* and *adapted* rubber springing for his bicycle. The significance of the Moulton bicycle was not just its novel F-frame design – suitable for both men and women and with extra load-carrying capacity – but the fashionable image and revival of interest it gave to cycling and cycle design (Figure 1.5a). *Which?* tested a Mark 1 Moulton bicycle in 1964 and concluded, 'For utilitarian purposes, and as a machine that can be shared by several people ... the Moulton bicycle has advantages which make it better value for money than a comparable bicycle of conventional design' (Consumers' Association, 1964).

Although they sold well and evolved through several models, Moulton's manufacturer, Raleigh, eventually stopped making his bicycles in 1974 after the company had developed a cheaper small-wheel bicycle without suspension. In 1977 Moulton therefore decided to develop a new small-wheel bicycle that performed as well as a good conventional bike. He eventually created a multi-tubular space frame design that, as well as being light and strong, was very stiff. The 'advanced engineering' Moulton AM series was first launched in 1983, aimed at the discerning cyclist at the top end of the market (Figure 1.5b).

10 Patterns of innovation

FIGURE 1.5(a) Alex Moulton with three models of his small-wheel bicycle with rubber suspension, 1967

FIGURE 1.5(b) Alex Moulton riding his 'advanced engineering' space frame bicycle, 1983

Design for different users and markets

The renewed interest in cycling, arguably stimulated by Moulton's innovation, led to a wide variety of new cycle designs from the 1970s onwards. The new designs included many types of folding bicycle, such as the highly successful

Brompton bicycle (Figure 1.6), typically used for commuting with other modes of transport. More radical innovations include recumbent bicycles, in which the rider sits in a reclining position to reduce air resistance, and streamlined human-powered vehicles (HPVs) for breaking speed records (Figure 1.7). It was the creative *combination* of his experience in designing both racing bicycles and HPVs that inspired Mike Burrows, a British cycle designer, to conceive and make an innovative racing bicycle with an aerodynamic carbon fibre frame (Figure 1.8). Burrows' prototypes were subsequently developed into the LotusSport bicycle, which was used to win the 4,000 metres individual pursuit event at the 1992 Olympics (Candy and Edmonds, 1996).

However, probably the most significant innovation was the mountain bike. Mountain bikes had their origins in the heavy-duty machines with balloon tyres built by Californian enthusiasts in the late 1970s for downhill racing and off-road riding. In 1981 the first mountain bikes began to appear on the mass

FIGURE 1.6 Brompton folding bicycle

FIGURE 1.7 Streamlined human-powered vehicle (HPV)

FIGURE 1.8 Mike Burrows' Mark 1 racing bicycle with aerodynamic monocoque (single shell) carbon fibre frame, 1985

market in the USA, reaching the UK in 1984. Mountain bikes also stimulated innovations in components and materials. In particular, Japanese manufacturers were quick to spot the commercial potential of these new designs and developed new components, such as disc brakes and multi-speed gears that could be changed without the rider losing power. By 1990 mountain bikes, often with rear and/or front suspension systems, accounted for over half of the bicycles sold in industrialised countries, reviving the fortunes of the cycle industry and helping to associate cycling with freedom, fitness and affluence.

As well as the mountain bike, diamond-frame bicycles are now produced in a variety of models for different uses and consumers. These range from mass-produced, utility bicycles with ordinary steel frames costing under £100 and 'hybrid' bicycles (Figure 1.9) to road and racing machines with carbon fibre frames costing thousands of pounds (Figure 1.10).

When testing bicycles, *Which?* focuses on the most popular types – utility, commuter and mountain bicycles – and rates them for price; specification; ride, gearing and braking performance; construction quality; and ease of use (Which? Ltd., 2005). *Which?* does not judge the aesthetics of the products it tests, although in the case of bicycles their look and a fashionable brand are very important to most cycle buyers.

Electrically assisted cycles, for people who wish to reduce the effort involved in cycling, are a new category beginning to be produced in an increasing variety of designs and sold in increasing numbers outside their main markets in Japan and China (Figure 1.11). Even the famous French designer, Philippe Starck, has designed a range of electrically-assisted bicycles, each suited to a different terrain (Dezeen, 2014).

FIGURE 1.9 Hybrid bicycle, a general purpose or commuting bicycle that combines features of mountain, road and touring bikes, c. 2005

FIGURE 1.10 Bianchi Sempre road and racing bicycle with aerodynamic carbon fibre frame, 2012

FIGURE 1.11 Smart electrically-assisted bicycle

Environmental impacts

While concern about the impacts on the environment of other consumer products has grown, the impact of making, maintaining and disposing of bicycles has not been considered a significant problem. Rather, as the mode of transport with the lowest environmental impacts, the bicycle itself has become a symbol of environmental sustainability. One study, for example, calculated that over its life cycle from manufacture through use and maintenance, a bicycle uses only about 7% of the energy and produces less than 10% of the greenhouse gas (GHG) emissions per passenger kilometre than either a medium-sized petrol car or an averagely-occupied bus (Dave, 2010). Electrically-assisted bicycles produce about double the greenhouse gas emissions per passenger kilometre than ordinary pedal bicycles, but still only a fraction of the GHG emissions of a medium car (Del Duce, 2011).

Social influences and impacts

The evolution of the bicycle has been the subject of several other studies. For example, van Nierop *et al.* (1997) plotted the rise and fall in popularity of different cycle designs according to their fitness in meeting the needs of different users and a changing social context. Their 'fitness landscape' shows a peak in popularity of the Ordinary (high-wheel or penny-farthing) around 1873 and its virtual disappearance by 1893, tricycles peaking around 1875 and almost gone by 1900, while the rear chain-driven diamond-frame bicycle steadily increased in popularity from about 1884 and had wiped out most rival designs by 1900.

Such work indicates that the development of an artefact such as a bicycle does not follow an inevitable path of technical progress. Instead, as some sociologists of technology argue, the design of technological artefacts is 'shaped' by social factors, especially the preferences and influence of relevant

> social groups (RSGs). In support of this 'social construction of technology' (SCOT) theory, a paper by Pinch and Bijker (1984) argued that Ordinary (high-wheel) bicycles only succeeded because they met the needs of a particular RSG – young sporting male cyclists – who viewed them as 'macho machines' that symbolised speed and daring. For other RSGs – women and older men – high-wheel bicycles were considered dangerous and difficult to ride. The unsuitability of the high-wheel design for these RSGs stimulated inventors and manufacturers to design safer cycles, resulting in the classic diamond-frame bicycle. But the diamond-frame design only finally displaced rival designs after it had been fitted with pneumatic tyres, which enabled it to simultaneously meet the needs of *all* RSGs – of women and older men for safety and comfort, and racing cyclists for speed.

Theories of technological and design change

I included the bicycle case study because its pattern of innovation and evolution provided a framework for researching the book's other product case studies. It shows a pattern of early experimental design divergence resulting in convergence on a dominant design followed by a second period of divergent innovative design. At each phase – even the dominant design phase – many different designs are produced and marketed. Explanations for this pattern of innovation and the commercial success and failure of particular designs may be found in the literature of several disciplines, including the history of technology, design studies, innovation theory, social studies of science and business management. There is not space in this book to discuss the large volume of literature on technical and design change, as reviewed for example by Muir-Wood (2010). So only a few concepts to guide the development and analysis of the case studies will be introduced here.

The fluid or experimental phase

The idea that products evolve through phases, has been proposed by several innovation theorists, notably in influential publications by Utterback and Abernathy (1975) and Abernathy (1978), based on a study of the evolution of the internal combustion engine. What Utterback and Abernathy call the Stage I or 'fluid' (i.e. early experimental) phase starts with the invention and innovation of a new product class (e.g. the motor car, bicycle, television) which leads to the creation of a great diversity of designs as inventors, designers and manufacturers compete to create products that function better. Henry Petroski (1993), an engineer and historian of technology, has therefore argued, based on his studies of everyday objects from paper clips to aluminium cans, that 'the form of made things is always subject to change in response to their real or perceived shortcomings'. Another historian of technology, George Basalla (1988), demonstrates that these diverse designs comprise different configurations and combinations of available materials, components and technologies, which are also evolving. Subsequently, Basalla argues, which product designs succeed and which disappear depends, not just on how well they function but, amongst other factors, on 'the power of cultural values, fashions and fads'. Social scientists such as Bijker (1995) point also to the role of the preferences, influence and power of relevant social groups in the creation and selection of commercially successful products.

The transitional or dominant design phase

The next phase is what Utterback and Abernathy (1975) and Abernathy (1978) call the Stage II or 'transitional' phase. They observe that, unless they serve a separate niche market, manufacturers converge on a dominant design which functions satisfactorily and fits consumers' expectations. Attention shifts from product innovation to process improvements as manufacturers compete to produce the dominant design more efficiently and cheaply. For this phase Giovanni Dosi (1982) proposed the concept of a 'technological paradigm'. A technological paradigm – analogous to Thomas Kuhn's (1962) concept of a scientific paradigm – defines what technical problems designers, engineers and manufacturers consider need addressing and the ways of working and the knowledge required for their solution. Within the technical viewpoint and knowledge constraints of the paradigm, product designs undergo incremental improvement along what Dosi calls a 'technological trajectory'. Radical innovations are therefore unlikely to be developed and introduced during the dominant design phase, because such innovations would require a new technological paradigm (like one of Kuhn's scientific revolutions) based on different knowledge and ways of thinking and working.

At the product level, therefore, instead of radical innovation there is an emphasis on incremental improvement and product differentiation as manufacturers compete for a share of the market. The result is a diversity of products which are variants on the dominant design. The products may have different specifications and features, be made from different materials, come in different colours and finishes and vary in reliability and durability. They are likely to be aimed at different consumer groups or market segments, cost different amounts to make and be sold at different prices. For example, Muir-Wood (2008) analysed *Which?* electric kettle Best Buys from 1971 to 1997 and found 85 different designs and models from 24 brands all heated by an immersed electric element but differentiated by shape (traditional and jug forms); corded and cordless models; body material (stainless steel, plastic, aluminium, etc.); and price.

As will be shown in the case study chapters of this book, not all dominant designs – washing machines, for example – display such variety. But even then there are many different models at different prices produced by different manufacturers. In this phase, with the basic technology and key components fixed, the contribution of industrial designers becomes increasingly important, especially for product styling and to improve usability. The design of consumer products is thus, to a greater or lesser extent, influenced by fashion. The design historian, Ferebee (1970), notes the evolution in consumer product designs from the flowing, streamlined styles of the 1930s and 1940s to the sharper, rectangular shapes of the 1950s and 1960s and then to the more rounded, sculptural styles of the 1970s onwards. Design historians also find that designers may continue to copy outdated forms, such as light-bulbs shaped like candles (Steadman, 2008), or push styles to non-functional extremes; for example, the fashion-driven tailfins on 1950s and 1960s American cars (Ferebee, 1970). A select few products become what have been called 'design classics'; long-lived products that evolve into a family of related designs, such as Sony's Walkman and Apple's iPod portable music players (Utterback et al., 2006).

During this phase of evolution, designing or redesigning their products for reduced environmental impacts may become one of the ways through which manufacturers compete and respond to environmental standards and regulations. There are many possible *ecodesign* strategies, as described, for example, by Brezet and van Hemel (1997). Products may be designed to be more energy or water efficient; use fewer, natural or renewable materials; avoid the use of hazardous or toxic substances; or be designed for longer life or for disassembly and recycling. If the strategy includes economic and social as well as environmental considerations, for instance, fair trade in its raw materials or good working conditions in the factories making it, the product may be considered to be a *sustainable* design, as discussed for example by Datschefski (2001) or Roy (2006). Ecodesign and sustainable design will be discussed in more detail in the final chapter of this book.

The innovative design phase

Although the introduction of radical innovations is highly constrained during the dominant design phase, inventors, designers and companies do not stop researching, inventing or creating novel concepts, designs and technologies. Sooner or later one or more radically innovative products will succeed in reaching the market and begin to be adopted by consumers. They may remain as niche products or, if successful, they may begin to displace or eventually replace the dominant design and its underlying technology. In this chapter I gave the example of the Moulton small-wheel bicycle, which challenged but did not replace the dominant classic diamond-frame bicycle.

Several scholars have researched this phase of innovation. Military metaphors are often used to describe how a radical innovation based on a new technology challenges an established technology and industry. For example, Foster (1988) writes about the innovation 'attacking' established businesses and Utterback (1994) writes about the new technology 'invading' the market. Both Foster and Utterback thus emphasise the difficulties for established companies operating within the dominant technological paradigm shifting their resources and efforts from incremental product improvement to radical innovation in order to respond to the invasion or attack. Christensen (1997) argues that radical innovations typically start as niche products created by new enterprises. These innovative products, compact fluorescent lamps, for example, may at first have an inferior performance to the dominant design produced by the established industry but nevertheless appeal to a few consumers. But as the innovative products are improved they may begin to be adopted more widely and so displace or replace the dominant design and technology – they then qualify as what Christensen calls a *disruptive innovation*. However, if the innovation depends on technologies, materials, components or infrastructures that are insufficiently developed, or its looks are too novel or unusual, it may fail to catch on and die. For example, some early electrically-assisted cycles, such as Sinclair's 1985 C5 tricycle and 1992 Zike bicycle, suffered from insufficiently developed batteries and motors and designs that were too novel for consumers, and so they were soon discontinued.

Disruptive innovations are significant because once the hold of the dominant design and the old technological paradigm has been disrupted, further radical innovations are likely to be developed and introduced. Hence a second phase of technological and design diversity emerges; for instance, the upsurge in new bicycle designs that followed the launch of the Moulton bicycle. The innovative products that appear during this phase may include ones with significantly lower environmental impacts; for example, light emitting diode (LED) lamps compared to incandescent and compact fluorescent light bulbs.

Helping consumers to choose from this great diversity of products is one of the main reasons for organisations like the Consumers' Association in the UK, *Stiftung Warentest* in Germany, Consumer Reports in the USA and the many other consumer and product testing organisations worldwide. Specialist magazines are another source of product tests and reviews, and since the rise of online shopping, consumers are also helped to choose products by the reviews posted by other consumers on shopping and other websites.

Practical lessons for product developers

The case of the invention, innovation and evolution of bicycles and the above discussion of patterns of innovation are not just of academic interest. They illustrate a number of general points and practical lessons for product designers, engineers, developers and managers. These general points and practical lessons will be taken up in more detail in the final chapter, making use of empirical evidence from the other five consumer product case studies and further use of the literature and theories concerning technological change and design evolution discussed in this chapter.

References

Abernathy, W. J. (1978) *The productivity dilemma*, Baltimore, MD, Johns Hopkins University Press.
Basalla, G. (1988) *The Evolution of Technology*, Cambridge, Cambridge University Press.

Bijker, W. E. (1995) *Of Bicycles, Bakelites, and Bulbs: Toward a theory of sociotechnical change*, Cambridge, MA, MIT Press.

Brezet, H. and van Hemel, C. (1997) *Ecodesign. A Promising Approach to Sustainable Production and Consumption*, Paris, United Nations Environment Programme.

Candy, L. and Edmonds, E. (1996) 'Creative design of the Lotus bicycle: implications for knowledge support systems research', *Design Studies*, vol. 17, no. 1, pp. 71–90.

Christensen, C. M. (1997) *The Innovator's Dilemma: When New Technologies Cause Great Firms to Fail*, Boston, MA, Harvard Business School Press.

Consumers' Association (1964) 'The Moulton bicycle', *Which?* August, pp. 246–247.

Consumers' Association (1995) 'Washing machines you can rely on', *Which?* January, pp. 50–51.

Dave, S. (2010) 'Life Cycle Assessment of transportation options for commuters', [Online] Unpublished paper, Massachusetts Institute of Technology (MIT), February. Available at http://files.meetup.com/1468133/LCAwhitepaper.pdf (accessed December 2014).

Datschefski, E. (2001) *The Total Beauty of Sustainable Products*, Crans-près-Céligny, Switzerland, RotoVision.

Del Duce, A. (2011) Life Cycle Assessment of conventional and electric bicycles, [Online] Eurobike 2011 Friedrichshafen, 2. September. Available at http://www.eurobike-show.com/eb-wAssets/daten/rahmenprogramm/pdf/LifeCycleAssessment_DelDuce_englisch.pdf (accessed December 2014).

Dezeen (2014) Philippe Starck launches MASS cycling collection at Eurobike, *Dezeen*, 27 August [Online] Available at http://www.dezeen.com/2014/08/27/philippe-starck-mass-electric-bicycle-cycling-accessories-eurobike/ (accessed December 2014).

Dosi, G. (1982) 'Technological paradigms and technological trajectories: A suggested interpretation of the determinants and directions of technical change', *Research Policy*, vol. 11, pp. 147–162.

Ferebee, A. (1970) *A History of Design from the Victorian Age to the Present*, New York, Van Nostrand Reinhold.

Foster, R. N. (1988) *Innovation: the Attacker's Advantage*, New York, Summit Books.

Kuhn, T. (1962) *The Structure of Scientific Revolutions*, Chicago, University of Chicago Press.

Muir-Wood, A. (2008) 'Kettles – Which', Unpublished Excel spreadsheet, Cambridge, 5 May.

Muir-Wood, A. (2010) 'The Nature of Change in Product Design. Integrating Aesthetic and Technical Perspectives', Unpublished PhD thesis, Cambridge, University of Cambridge, September.

Petroski, H. (1993) *Evolution of useful things*, New York, Alfred A. Knopf.

Pinch, T.J. and Bijker, W. E. (1984) 'The social construction of facts and artefacts: Or how the sociology of science and the sociology of technology might benefit each other', *Social Studies of Science*, vol. 14 (August) pp. 399–441.

Roy, R. (1980) Bicycles: invention and innovation, T263 *Design: Processes and Products* Units 5-7, The Open University, Milton Keynes.

Roy, R. (1994) 'The evolution of ecodesign', *Technovation*, vol. 14, no. 6, pp. 363–380.

Roy, R. (1997) 'Design for environment in practice – development of the Hoover New Wave washing machine range', *Journal of Sustainable Product Design*, Issue 1, April, pp. 36–43.

Roy, R. (1999) 'Design and Marketing Greener Products: the Hoover Case', in Charter, M. and Polonsky, M.J. (eds) *Greener Marketing* (2nd edition), Sheffield, Greenleaf Publishing, pp. 126–144.

Roy, R. (2006) 'Products: new product development and sustainable design', T307 *Innovation: designing for a sustainable future* Block 3, Milton Keynes, The Open University.

Roy, R. and Tovey, M. (2012) 'Bicycle design: creativity and innovation', in Tovey, M. (ed.) *Design for Transport. A User-Centred Approach to Vehicle Design and Travel*, Farnham, UK, Gower.

Smith, M.T., Roy, R. and Potter, S. (1996) The Commercial Impacts of Green Product Development, Design Innovation Group, Report DIG-05, Milton Keynes, The Open University, July (ISBN 0 7492 8831 0) [Online]. Available at http://design.open.ac.uk/documents/GreenDesRptFinalPix3_000.pdf (accessed February 2015).

Starley, W.K. (1898) 'The evolution of the bicycle', *Journal of the Royal Society of Arts*, vol. 46, May 20, pp. 601–616.

Steadman, P. (2008) *The Evolution of Designs: Biological Analogy in Architecture and the Applied Arts* (Revised edn) New York, Routledge.

Utterback, J. M. and Abernathy, W. J. (1975) 'A dynamic model of process and product innovation', *Omega*, vol. 3, no. 6, pp. 639–656.

Utterback, J.M. (1994) *Mastering the Dynamics of Innovation*, Boston, MA, Harvard Business School Press.

Utterback, J.M., Vedin, B., Alvarez, E., Ekman, S., Sanderson, S.W., Tether, B. and Verganti, R. (2006) *Design-inspired innovation*, Singapore: World Scientific.

van Nierop, O.A., Blankendaal, A.C. and Overbeeke, C.J. (1997) 'The evolution of the bicycle: a dynamic systems approach', *Journal of Design History*, vol. 10, no. 3, pp. 253–267.

Which? Ltd. (2005) 'Bicycles', *Which?* July, pp. 43–47.

2
WASHING MACHINES

The modern washing machine has probably done more than any other consumer product to relieve people – overwhelmingly women – from the arduous task of washing and drying clothes and other household textiles. It is basically a device for agitating the laundry in water containing a cleaning agent in order to remove dirt and stains, usually with an integral or separate device for removing much of the water from the washed laundry.

Before the invention, and then widespread adoption, of mechanical and electric washing machines, clothes and textiles were only washed occasionally, using a variety of traditional methods. These included beating on stones, washing in rivers and ponds or in washhouses or communal washing sinks. By the nineteenth century the laundry was more often done by agitating it in wooden or metal tubs with a washing bat, a pronged 'dolly', or a 'posser' with a cone end, or scrubbing on washboards. Large houses would have a laundry, but smaller homes had to wash in the kitchen or the yard. Before indoor plumbing, water often had to be carried some distance. Rainwater was sometimes stored for washing, and rural communities could still use streams. Soaps made from wood ash and animal fat were used to help remove dirt and stains before industrially-manufactured soap became more widely available. The water used was either cold or heated over a fire. The laundry was dried by wringing by hand, using a wringing net or hand mangle and then hanging on bushes, on clothes horses in front of a fire, or on pulleys or lines. All this was very hard and time-consuming work. Women, whether housewives, servants or washerwomen, often had to devote one or more days to the laundry – soaking, heating water, hand washing with or without aids like a dolly or wash board, then rinsing, wringing, drying and ironing.

The first electrically-powered washing machines only appeared at the beginning of the twentieth century, but it was not until the 1950s that washing machines, often with wringers or spin dryers, began to be found in many households in industrialised countries and the chore of washing day was eased. Now the majority of households in industrialised countries have a washing machine. The proportion for the UK in 2010 was 96%, while 56% also owned a tumble dryer (ONS, 2011) and these machines accounted for 11% of UK household electricity use and 9% of water consumption (DECC, 2013; EST, 2013).

The Swedish statistician, Hans Rosling, has worked out that of the world's population of about 7 billion, in 2010 only some 2 billion people in industrialised and newly industrialised countries had a washing machine, while the remaining 5 billion still washed by hand. For example, in 2011 about half of Brazilian households owned a washing machine while in India ownership was 14% in urban areas and 7% in rural areas (Spencer *et al.*, 2012). Rosling estimated that by 2050, 3 billion more people will have electricity and will want to own a washing machine, and argues that this will free those people from one of their hardest and most time consuming tasks (Rosling, 2010).

Washing machine technology and design

Manual washing machines

Before 1800 few people had seen, or even heard of, a washing machine, and before 1900 a machine for laundering could only be found in a few wealthy homes in rapidly industrialising countries. Nevertheless, in the eighteenth century, several attempts had been made to mechanise laundry work. For example, in Germany in 1766 Jacob Schaffer published the design for a wooden tub in which a pronged wooden paddle ('dolly') could be turned by hand (Figure 2.1a).

Other patented devices followed in the late 1700s and early 1800s, including various designs of 'washing mill' in which the laundry was agitated, rotated or tilted in a wooden container filled with soapy water via a hand-operated mechanism. An example was the machine promoted by Edward Beetham in England in the 1790s in order to save on the cost of servants and to wash clothes with less wear and tear (Figure 2.1b shows a large version designed for use on ships).

Various improved hand-operated washing machines were produced in the nineteenth century, aimed mainly at upper- and middle-class households with servants, including by the mid-1800s machines with attached mangles or wringers (Figures 2.2a and 2.2b). Washing soap in bars and flakes and other washing ingredients were developed and became more affordable.

Early electric washing machines

With the development of small electric motors and electrical distribution systems at the beginning of the twentieth century (for more details see Chapter 3, Lamps and lighting), the first electrically-powered washing machines were invented and introduced. The first to be mass produced, although not quite the first to be designed, was the 1907 Thor washing machine made by the Hurley Machine

FIGURE 2.1a Jacob Schaffer's washing machine, Germany, 1766

FIGURE 2.1b Edward Beetham's washing mill, England, 1790s

FIGURE 2.2a Improved Doty hand-operated squeezer washing machine, New York, c. 1870

FIGURE 2.2b 'Pioneer Masher' manual washing machine and mangle, Accrington, England, 1903

Company of Chicago. It had a cylindrical wooden tub, driven by an electric motor via pulleys and gears, in which the laundry was tumbled (Maxwell, 2009). A chain-driven wringer was also provided (Figure 2.3).

Like a modern front loading washing machine, the Thor had a horizontal-axis tub (which was top loaded), but there was no water heater and the motor was

24 Washing machines

FIGURE 2.3 Thor washing machine, Chicago, c. 1907, designed by Alva J. Fisher and patented 1910, was one of the first electric washing machines. The exposed electric motor under the tub sometimes caused hazardous electric shocks.

FIGURE 2.4 Electric washing machine with a powered wringer, made by Beatty Bros of Canada, c. 1920. The wooden tub with a four-legged rotating wooden dolly inside is like those of earlier manual machines. The exposed motor could be hazardous

mounted externally, which if it got wet could cause hazardous electric shocks. However, many washing machines up to the 1920s were similar to manual machines with a fixed vertical wooden tub, but with the agitating bars or rotating dolly driven by an external electric motor (Figure 2.4). This is an example of the 'horseless carriage' approach to design in which a traditional product is mechanised without changing its basic architecture (Robertson, 1994).

The Lee Maxwell Washing Machine Museum, Colorado, USA has a huge collection of old washing machines from early manual types to recent automatics. It displays an astonishing variety of designs, classified by the method of agitating the laundry in water, including rotating dollies; tilting tubs; tumbling drums; various types of propeller; gyrator or paddle; plus vacuum and vibrating methods.

In the 1920s new washing machine designs were introduced, including ones with agitators or paddles similar to those used on modern top-loading machines. Wooden tubs were increasingly replaced by metal ones; the motor was sometimes enclosed and drove the agitator and wringer through belts and gears. By the 1930s the designs were becoming less like 'Heath-Robinson' contraptions and more like domestic appliances, typically comprising a top-loading metal tub on legs with the drive motor underneath and a hand or motor-driven wringer on top.

Dominant designs

Modern washing machines with white enamelled steel cabinets enclosing the motor, tub and an internal electric water heater began to appear just before and after the Second World War, such as the Maytag top-loading machine of 1939 (Figure 2.5). However, as early as 1937 Bendix had launched its first fully automatic front-loading machine aimed at wealthy American households. It was granted a patent in 1939 and was embodied in different designs (Figures 2.6a and 2.6b).

Hoover introduced its first washing machine made in the UK in 1948, a single tub top-loader with an impellor agitator. At that time a washing machine was still a luxury and only about 4% of British households owned one. Most people still hand washed the laundry, often with the aid of a washboard, an electric boiler for towels, etc. and a wringer for drying.

Although a variety of types continued to be produced after the Second World War, washing machines were beginning to converge on two dominant configurations: top-loading designs with various types of vertical agitator and front-loading designs with horizontal rotating drums (Maxwell, 2013). The cabinet forms also began to converge from a variety of cylindrical and sink-shaped machines on legs, often with attached wringers, and reminiscent of earlier designs, into the familiar white boxes of today.

By 1958, an early *Which?* report (Consumers' Association, 1958) noted that there were over 50 washing machines on the UK market and provided test results for six machines. These were all single-tub, top-loading machines, either with

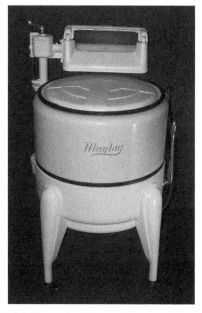

FIGURE 2.5 Maytag Model 18 top-loading washing machine with wringer, USA, 1939

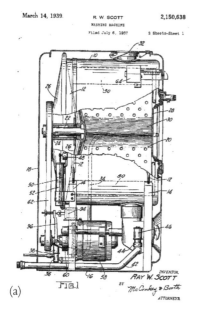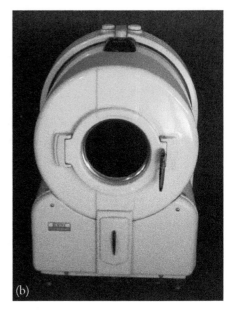

FIGURE 2.6 The first mass-produced fully automatic washing machine, the Bendix Home Laundry was first launched in 1937
(a) Scott's automatic washing machine patent
(b) Bendix automatic, c. 1945

a rotating impellor or 'pulsator' that swirled the laundry in one direction or an agitator or propeller that moved the water and laundry to and fro. The machines had a pump to drain the water into a bucket or sink, plus an electric water heater, although many users filled their machine with hot water as using the machine's heater took a long time. In late 1950s Britain, buying one of these washing machines, costing £60 – £70 (about £1,250 to £1,500 today), was a major expenditure, so *Which?* compared the cost of buying and running a machine with using a launderette or a laundry and concluded that buying a machine was easily the lowest-cost option. The machines tested all had a power wringer to help dry the laundry, which was then hung up to dry or maybe dried on an electric 'airer', since in 1958 spin driers, and certainly tumble driers, were only beginning to reach the UK market.

A different design which helped to spread washing machines to many more households was the twin tub; a machine that contained a separate washer and spin dryer. Twin tub designs had existed since the 1920s, but started to become popular in British homes in the late 1950s. Hoover launched its first Hoovermatic twin tub in 1957, but the product that introduced many British households to the twin tub in the late 1950s and early 1960s was the Electromatic/Starmatic made by Rolls Razor. These machines were sold direct from the manufacturer at £54 without a heater and £72 with a heater, and due to their cost (about £1,000 to £1,200 today) were mainly sold on the 'never-never' – weekly hire purchase.

However, a 1960 *Which?* report said, 'in the past year or two the trend has been away from separate machines towards a unit which both washes and part dries clothes'. The report identified three types: twin tubs, which required the user to set controls, load the laundry and transfer, rinse and dry it in the spin drier; semi-automatic single tubs, either top- or front-loading, which required the user to advance the controls and wring the laundry; and automatic front-loading machines, which only required the user to load the laundry and set the controls. By the mid-1960s twin tubs began to have automatic wash programmes, only requiring user intervention for spinning and rinsing; while automatics offered four or more wash programmes (Consumers' Association, 1960, 1964a, 1964b, 1966).

Which? reported its last major test of twin tubs in August 1981 (although you can still buy a twin tub in 2015 for the advantages these machines offer – flexible loading, shorter wash times, high spin speeds and the ability to reuse the hot water for several washes – Figure 2.7). By the early 1980s twin tubs had increasingly been superseded in Britain by fully automatic machines (Consumers' Association, 1975, 1976, 1981). Most of these were front-loaders with a porthole door of the familiar design pioneered by Bendix in 1937 (Figures 2.8a and 2.8b). But there were variants, such as the 1961 Hoover Keymatic, which had a sloping drop-down front door, an inclined drum with a rotating disc impeller in the base, and a plastic key that the user inserted to set one of eight wash programmes (Figure 2.8c). Other variants were top-loading automatics, mostly with vertical-axis drums (Figure 2.8d) and combined washers and tumble driers.

The automatic washing machine had thus become the dominant design. The horizontal axis front-loading type is the most common design in Europe, the Middle East, Asia and many other parts of the world. The vertical axis top-loading washer is most popular in the US, Canada, Australia, New Zealand, Latin America and Japan, although upmarket washing machines in these areas tend to be

FIGURE 2.7 Modern Indian twin tub washing machine, 2015

FIGURE 2.8 Some automatic washing machines on UK market in 1966
(a) Bendix horizontal axis front-loading
(b) English Electric horizontal axis front-loading
(c) Hoover Keymatic inclined axis front-loading
(d) Hotpoint vertical axis top-loading

front-loaders. Low-cost simplified top-loaders are made for markets in Asia, Africa, and other less-developed parts of the world (Wikipedia, 2013).

The basic design of automatic washing machines remains unchanged, but there have been numerous improvements since the early machines. These focused initially on increasing spin speed, reducing manufacturing costs and improving reliability. Typical spin speeds on automatics doubled from about 500 rpm to around 1000 rpm by the early 1980s and then increased to 1200 – 1500 rpm.

Higher spin speeds reduced line drying times or the energy use and cost of using a tumble drier when these started entering the UK market from the 1970s. The controls on many front-loaders moved from the top (e.g. Figures 2.8a, 2.8b) to the front to allow machines to fit under a worktop. The design of the machines then gradually evolved, for example, the square boxes of the 1990s (Figure 2.9) became slightly rounded at the front and the drums increased in capacity. The electromechanical controls on all but the lowest cost products were replaced by microelectronic controls with digital displays, offering a wide range of wash programmes (e.g. Figures 2.12 and 2.13).

In this mature phase of washing machine technology with two basic designs, manufacturers competed on cost and reliability within market segments differentiated by maximum spin speeds, sophistication of controls and styling. As concern about the environmental impacts of appliances grew, designers and manufacturers turned to reducing the energy, water and detergent use of their machines as a way of gaining a competitive advantage.

Environmental impacts and regulation

The oil crisis of the early 1970s raised the issue of saving energy in the home, but the environmental impacts of home laundry did not become of serious concern until later. This concern was stimulated when in the late 1980s the EU proposed an ecolabelling scheme, with washing machines and dishwashers as the first products to be assessed.

Life cycle assessment

To help develop the criteria for the Ecolabel, a life cycle assessment (LCA) study of washing machines was conducted in 1991. This calculated the environmental impacts of washing machines over their life cycle from production (including materials consumption and manufacture of components) through distribution and

FIGURE 2.9 Some *Which?* Best Buy automatic washing machines on UK market in 1994. Left to right: Candy Aquaviva; Zanussi washer drier; Zanussi Jetsystem

use to disposal. The LCA showed that over 90% of the environmental impacts occur during the use phase of the life cycle as the result of the energy used by the machine, its consumption of water and detergent, and from the need to treat its waste water (Figure 2.10). The study concluded that the most important criteria for the award of a washing machine Ecolabel were low energy, water and detergent consumption (Durrant et al., 1991).

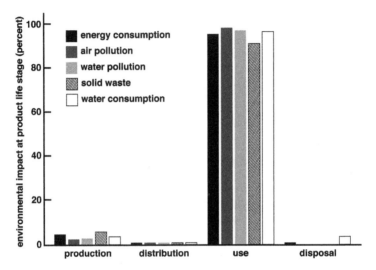

FIGURE 2.10 Percentage contributions of the production, distribution, use and disposal phases of a washing machine's life cycle to its environmental impacts, excluding detergent manufacture and distribution. (Adapted from Roy (2006) p. 132 and Durrant et al. (1991) p. 37, Figure 2.3)

The energy use involved in detergent manufacture was not included in the LCA, but German data indicated that this was the second most important factor in the environmental impacts of home laundry, increasing the energy consumed at the use stage by about 50%, as well as producing significant pollution and waste (Durrant et al., 1991).

Another LCA carried out for the Australian Consumers' Association (Deni Greene Consulting, 1992) confirmed the findings of the EU study that the main environmental impacts of washing machines occur at the use phase. It also confirmed that detergent manufacture and packaging was the second most important source of impacts, increasing the use impacts by about a third to half. The LCA also showed that top-loaders produced about twice the impacts of front-loaders for a warm wash. But for a cold wash (common in Australia and some other countries) the impacts of using the machine fell by 50% to 75% and the differences between top- and front-loaders were greatly reduced.

Consumers' Association tests

As early as 1960, well before environmental issues became prominent, a *Which?* test report included the water consumption and the energy required to heat the water in automatic, semi-automatic and twin-tub washing machines. It found that for 4–5 kg laundry, 27–100 litres of water were used for washing and rinsing. About 3 kWh was used to heat the water to 85°C, and about 2 kWh to 65°C, with no big differences between the three types of machine (Consumers' Association, 1960).

By 1979 *Which?* found that the equivalent energy use (for the whole process, not just water heating) for front-loading automatics had fallen to 1.7–2.6 kWh for a 4–5 kg hot (95°C) wash, but water consumption at 45–150 litres had apparently increased. However, by then detergents had improved to provide effective cleaning at lower temperatures, hence the machines had wash programmes that used much less energy: from 0.5–1 kWh per wash at 50°C (Consumers' Association, 1979).

The automatic top loaders tested used considerably more electricity and water, at around 3 kWh and 170–180 litres of water for a hot wash, than the efficient models of front loaders (Consumers' Association, 1979). This is because a front-loader, in which the laundry tumbles through the water at the bottom of a horizontal tub, uses less water, and hence heating energy, than a top-loader in which the washing is immersed in a vertical tub and moved to and fro by an agitator.

By the time of the 1991 EU ecolabelling study 75% of the washing machines sold in the EU were automatic front-loaders and 25% top-loaders. But the environmental performance of washing machines did not seem to have improved since the 1979 *Which?* Report; using 1.8–3.2 kWh and 60–125 litres water for a 4–5 kg hot (95°C) wash (Durrant *et al.*, 1991). A 1991 *Which?* report gave similar results, and also showed that a 50°C wash was considerably more energy- and water-efficient than a 95°C one, mostly using 1kWh or less per wash. Another efficiency measure was the amount of detergent that is wasted by being washed into the sump, which ranged from less than 10% to over 20% (Consumers' Association, 1991).

From the early 1990s *Which?* stopped reporting energy use in units of electricity, but instead provided annual running costs and sometimes star ratings for energy use, as that information was considered of more use to consumers. But other sources showed that at that time the average energy use of European washing machines ranged from 0.9–2.2 kWh per 5 kg load at 60° C with an average of 1.3 kWh (GEA, 1995).

Environmental labelling

Environmental labelling is one of the regulatory factors that have resulted in a reduction in the environmental impacts of consumer products. The German Blue Angel introduced in 1978 was the first voluntary Ecolabel, followed by the Nordic Swan and US Green Seal Ecolabels in 1989. The EU's voluntary Ecolabel scheme was announced in 1990 and introduced for washing machines in 1993, but superseded in the EU by mandatory Energy Labels.

32 Washing machines

The first product to be awarded an Ecolabel was a washing machine designed by Hoover in Britain, the development of which provides useful lessons about how one manufacturer responded to the pressures for more environmentally-friendly products.

CASE STUDY: THE HOOVER NEW WAVE WASHING MACHINE

The New Wave project began in the late 1980s when Hoover had a range of ageing washing machine designs, which were losing market share. The company had already decided to introduce a new method of making the cabinet by mechanically joining pre-coated steel panels and to replace the welded steel outer tub with moulded plastic. There was therefore an opportunity to develop a new, more up-market product range made by new manufacturing processes.

Hoover knew that consumers were increasingly demanding 'greener' products, that the EU was planning to introduce an Ecolabel for washing machines, and that competitors had already introduced water- and energy-saving machines. Zanussi, for example, launched its 'Jetsystem' range in the mid-1980s in which the water is pumped to the top of the machine and sprayed down onto the clothes instead of wetting them in the drum, thus saving water and energy.

Hoover therefore included environmental performance in the specification for its new range, championed by Hoover's Director of Engineering Development. The company realised that reductions in environmental impacts would depend mainly on minimising water, energy and detergent consumption in use. To achieve this, Hoover's researchers and engineers developed an innovative wash process. This involved filling the machine so that the laundry acted as a filter to reduce detergent lost into the sump, then perforated 'spray paddles' in the drum scooped up water and showered it over the clothes. New electronic controls also provided slow spins during the wash to wet the laundry using less water.

Prototypes then had to be developed into a production design, with industrial design of the cabinet and controls receiving considerable attention (Figure 2.11a). The 1991 LCA for the EU Ecolabel confirmed Hoover's focus on reducing water, energy and detergent consumption, but before the Ecolabel criteria were finalised in 1992 the company had also decided to address the impacts arising from production, distribution and disposal of the machine; for example, ensuring manufacturing energy and pollution were reduced and the machine was designed for disassembly and recycling.

The New Wave range was launched in February 1993 and was awarded an EU Ecolabel in November 1993 for exceeding set criteria for energy, water and detergent use per kg of washing and for its washing and rinsing performance (Figure 2.11b). (The New Wave used just 0.86 kWh at 60°C and 0.4 kWh at 40°C and 68 litres of water with no detergent loss for a 4.5 kg load.) *Which?* reported that the New Wave had the lowest running cost of the 19 machines it tested (Consumers' Association, 1994). Although it did not sell very well in the UK, Hoover found that the Ecolabel was an important factor in selling the New Wave, especially in the more environmentally conscious German, Danish and Austrian markets.

However, by 1997 the Hoover New Wave and its successor range were still the only appliances to have the Ecolabel. The main reason was that from April 1996 washing machines were subject to mandatory EU energy labelling and other manufacturers did not consider the voluntary Ecolabel worth its cost. The Energy Label for washing machines initially gave a ranking for energy use per kg laundry and for washing and spin drying performance, plus information on water consumption. The New Wave range achieved a B rating for energy efficiency and a C for wash performance (on the original A to G scale), while by then some other manufacturers' machines had higher ratings.

Hoover's new owners from 1995, the Italian Candy Group, introduced new models in 1998 that used an updated energy- and water-saving wash process. The cabinets were made from welded and powder-coated steel, replacing the innovative process used for New Wave, so as to be compatible with the Group's other factories. The new Hoover 'Quattro Easy Logic AA' machine achieved A ratings for energy and washing performance on the Energy Label, plus a low water consumption. Hoover therefore decided not to apply for an Ecolabel for its new range and the Ecolabel for washing machines was later withdrawn, although retained for detergents.

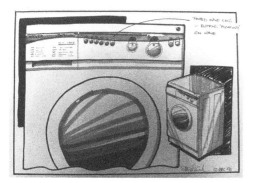 

FIGURE 2.11a Industrial design sketch of the Hoover New Wave washing machine controls

FIGURE 2.11b The New Wave range was awarded the first EU Ecolabel

34 Washing machines

No washing machine manufacturer could afford to have poor ratings on the Energy Label and all soon had models rated A or B. Thus in 2010 the EU washing machine Energy Label was revised with upgraded ratings from A+++ to D based on annual energy use per kg of laundry (Figure 2.12a), but omitting the wash performance rating.

In 2014 one of the most energy- and water-efficient front-loading machines from Bosch used 0.8 kWh and 40 litres water per 8 kg of laundry for a mix of 40°C and 60°C washes, earning it an A+++ energy label and a *Which?* Best Buy (Which? Ltd., 2013a). This machine (Figure 2.12b) uses about two-thirds of the energy and, by automatically adjusting the water use to the load, a third of the water per kg of washing when compared to the New Wave of 20 years earlier.

While Europe moved to high-efficiency washing machines, American consumers have traditionally favoured big top-loaders that use large amounts of energy, water and detergent. However, in 1992 the US Environmental Protection Agency introduced the voluntary Energy Star label programme, which applied to washing machines from 1997. Subsequently energy- and water-efficient washing machines became more widely available in the US. According to the EPA, Energy Star certified clothes washers on average use about 20% less energy and 35% less water than standard machines. Although the most energy efficient models are front-loaders, certified top-loaders are also available. The top-loading models save water and energy by flipping or spinning clothes through a stream of water instead of using a conventional agitator in a full tub of water. They also rinse clothes with repeated high-pressure spraying instead of soaking them in a full tub of water. Although most washing machines in American homes are still old inefficient models, by 2011 over 60% of new machines sold were Energy Star certified (EPA, 2013).

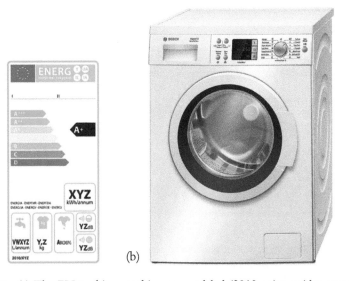

FIGURE 2.12 (a) The EU washing machine energy label (2010 on) provides energy and spinning efficiency ratings, annual electricity and water use and noise output; (b) Bosch washing machine with an A+++ energy label, 2014

Laundry innovation

Washing machine innovations

As noted earlier, while the basic design of most automatic washing machines has not changed since the original 1937 Bendix, there have been numerous innovations and improvements. Most of these aim to maintain or enhance wash performance while saving energy, water and/or detergent. Major manufacturers have introduced increasingly sophisticated controls on their mainstream products. For example, controls and sensors can match the amounts of water and detergent and number of rinses according to the wash load, and even to the level and type of dirt. This is part of a general trend of using IT to make electrical and mechanical products more 'intelligent'.

There has been active R&D on washing machine technology for many years. Many patents exist and there are several designs at prototype stage, or in production, that can further reduce water and detergent consumption. Innovations that have been tried in production machines include machines that electrolyse the water to produce active oxygen and hypochlorite ions to clean and deodorise clothes, and machines that wash with the aid of air bubbles. A design that used electrolysing technologies to clean lightly soiled clothes without the need for detergent was launched by Sanyo in Japan in 2002. More recently, Samsung Ecobubble™ machines mix detergent with water and air before pumping it all into the drum so the detergent will penetrate laundry quickly and efficiently. *Which?* (2014) gave the Ecobubble an excellent rating for cleaning at low temperatures and a Best Buy status (Figure 2.13a).

Dyson Appliances, which made its name with its bagless cyclonic floor cleaners (see Chapter 5), developed an innovative washing machine with two drums that rotated in opposite directions. Dyson asked his engineers to experiment with every imaginable way of washing, including hand washing, soaking, scrubbing, squeezing, water jets, ultrasound and micro-waves. The outcome was the Contrarotator™, launched in 2000. It was designed to replicate hand washing, which Dyson engineers found got clothes cleaner faster, by using contrarotating drums that agitated the laundry very thoroughly and vigorously (Figure 2.13b). *Which?* tested a later model of the machine, the Dyson CR02, in 2005 and found it the best and fastest machine for cleaning (Which? Ltd., 2005). However, the Contrarotator was launched at £1,000 and the CR02 sold at £800, £300 more than high-performing conventional machines. Because of its technical complexity and despite its high price, each of the relatively few Contrarotators sold lost money, so Dyson discontinued manufacture in 2005.

Another innovative, but also commercially failed, design was the Monotub Titan launched in 2001 at £650. This was a front-loader with a very large drum that contained a removable plastic washing basket. The drum was mounted at an angle, which meant the wash cycle could be paused to allow items to be added or removed. *Which?* tested the machine and reported that it 'made horrible noises and

FIGURE 2.13a Samsung Ecobubble™ washing machine, 2013

FIGURE 2.13b Dyson Contrarotator™ washing machine with two drums that rotated in opposite directions to simulate hand-washing, 2000–2005

jumped violently' when used (Which? Ltd., 2002). The company sold only very few Titans and went into liquidation in late 2002, with the intellectual property rights sold back to its inventor.

Such examples show the difficulty of innovating in an established industry that makes a dominant design that functions satisfactorily and can be sold at a competitive price. However, although radical innovation in such situations is very difficult, it is not impossible, as demonstrated by the success of the Dyson cyclonic vacuum cleaner that is discussed in detail in Chapter 5.

Developments in detergents

The drive to improve the performance and reduce the energy use of washing machines also depended on developments in detergents that removed stains and whitened at cooler temperatures. For example, by the late 1960s many synthetic and soap-based detergents contained optical brighteners (Consumers' Association, 1968).

The first 'biological' detergents – Radiant and Ariel – containing enzymes that broke down protein stains, such as blood, egg and milk, when washing below 65° C, were launched in Britain in 1969. Although *Which?* found that ordinary washing powders were generally as effective as these early biological powders (Consumers' Association, 1969), both types of detergent were steadily improved. More effective enzymes and other changes in detergent formulation allowed the two most common washing machine programme temperatures used in Europe to fall from 95°C for white cottons and 50°C for synthetics and coloured items, to 60° and 40°C and then to 40° and 30°C respectively. The latest machines have 15° or 20°C programmes, mainly intended for use with enzyme detergents.

The environmental impacts of washing machines also result from the discharge of the waste water containing detergent into rivers, lakes and estuaries. In particular, the phosphates, widely used in earlier detergents to soften water and increase the effectiveness of the cleaning agents, are plant nutrients and can accumulate in slow-moving water. This can contribute to the rapid growth of algae and other water plants which deplete oxygen and harm other aquatic life. Concern over this 'eutrophication' problem increased in the 1970s and 1980s and by the early 1990s many countries had banned, or introduced voluntary agreements to reduce the use of, phosphates, and other chemicals, especially zeolite, were substituted. By 1990 *Which?* found that about half the powder and liquid detergents on the UK market were phosphate free. These included brands of 'green' detergent such as Ecover and Ark which used plant- and mineral-based ingredients and were also free of synthetic chemicals such as water softeners, enzymes, brightening agents and perfume. However, tests showed that, at that time, green detergents were less effective at stain removal, although adding Ecover bleach to the powder made it perform as well as non-green products (Consumers' Association, 1990). By 2013 *Which?* found that Ecover performed as well as the non-biological detergents. Ecover Biological was also available, as detergents containing enzymes had become the most popular type, and in its 2013 formulation was as good at stain removal, but not as good at keeping fabrics bright white as the best alternatives (Which? Ltd., 2011, 2013b).

Drying the laundry

With the development of twin-tub machines and automatics, wringers and separate spin driers became redundant. Which? first tested three models of what were then called 'tumbler driers' in 1959, although tumble driers did not become popular in industrialised countries with a damp climate until the 1980s. By 2010 over half of UK and Dutch households owned a tumble drier, compared to 20% in Italy (Bush et al., 2013). However, tumble driers are greater energy consumers than washing machines. The latest vented or condenser tumble driers are likely to cost three to four times more per year to run than a new washing machine (Which? Ltd., 2013a, 2013c). So, consumers have been advised on cost and environmental grounds not to use tumble driers whenever possible. But using a tumble drier does reduce the relatively small electricity use, and the effort, of ironing.

More recently, to reduce their energy use, there have been innovations in tumble driers. These include gas-powered designs and tumble driers that have an electric heat pump to heat the air, which then reuses some of the heat released in the condenser; such designs halve annual running costs (Which? Ltd., 2013c).

Social influences and impacts

How laundry is carried out is deeply influenced by cultural norms and social habits, as well as by geographical factors such as climate. In the past, when all washing was done by hand, the laundry was done infrequently. Outer clothes made from wool and silk were not easily washable, but the wealthier a family the more clothes, especially underclothing, they possessed and so the longer they could stretch the time between washdays.

Washing as a weekly chore did not appear until the nineteenth century, when cleanliness became very important to the Victorian middle classes. Nevertheless, ways of saving on washing were encouraged, such as detachable collars and cuffs and patterned clothes that would not show the dirt.

Washing machines were promoted as a way of relieving women from drudgery. For example, advertisements for early electric washing machines showed a woman sitting at leisure while the machine did the work. Automatic washing machines were then sold as products that would make women's lives *even* easier. The Bendix Home Laundry advertisement of 1937 said: 'A thrilling announcement to every woman. At last a washless washday.' But in the 1950s, while many American women owned a washing machine, in Britain they were still only for wealthier households. In his memoire of growing up in a poor family in 1950s London, the former Labour Cabinet Minister Alan Johnson wrote:

> If [my mother] … were granted a wish for just one appliance … my guess is that she'd have chosen a washing machine. The effort involved in washing by hand, wringing … putting the clothes through the mangle, drying and ironing them … was hugely debilitating … and when she could afford it she'd send the washing to the 'bagwash', the predecessor of the launderette.
> (Johnson, 2013, p. 38)

Even as late as 1961 the Hoover Keymatic (Figure 2.8c) was advertised with the slogan 'washday just forget it'.

With laundering made easier and as more textiles became machine washable, standards of cleanliness and hygiene became stricter and so clothes and linen were washed increasingly frequently (Shove, 2003). By the time of the 2010 EU Energy Label, an average European washing machine was assumed to be used 220 times per year on full and part load. This agrees with other data that on average UK washing machines are used four to five times per week (EST, 2013), while in the US washing machines are used more frequently, nearly eight times per week with larger loads (Shove, 2004). So the automatic washing machine, with or without a tumble drier, on average has increased the weekly wash to an almost, or more than a, daily wash in industrialised countries. This coincided with changing social norms and habits that include the daily shower and more frequent changes of clothes, often washing them before they are dirty.

Apart from the frequency of laundering, consumer behaviours that affect its environmental impacts include the loading of the machine, the amount of detergent used, and the choice of washing temperature. Recent washing machines have increased capacity, but if they are only part loaded some of the energy and water saving may be lost, although having sensors that adjust the programme to the load reduces this problem. A large survey of UK consumers showed that 82% normally fully load their machine before use (EST, 2013), although consumers have been found to use too much detergent which then enters waste water. Only a few machines have automatic dosing to address the latter problem. As discussed previously, the temperature at which the laundry is washed has steadily fallen as detergents and machines have improved. But some users still use high temperature programmes from habit or the belief that this is essential for hygiene, even though high temperatures are only needed to avoid infection in vulnerable households. Another deterrent to cooler washes is that they usually take longer. The above-mentioned survey, however, showed that 70% of people normally washed at 40°C and 24% at 30°C, with only 6% still using a 60°C programme (EST, 2013).

Despite the introduction of cold water detergents, cold water washing is unusual in much of Europe and the US. But it is common in Japan and is normal in places where laundry can be dried and bleached in the sun. A cross-cultural survey of laundering by middle-class households in the UK, Brazil and India by Spencer *et al*. (2012) found that cold water washing is standard practice in Brazil and common in India, but rare in the UK. The Brazilian households washed once or twice per week, while in India laundry was done two to three times per week, more often when family members visited and to wash sweaty clothes. In both countries middle-class women had servants to help with the laundry and there were spaces in the home dedicated to laundry facilities, while relatively few UK households had home help or possessed a laundry or utility room.

Future developments

Although there is a well-established system for doing the laundry – using automatic domestic washing machines, and increasingly tumble driers – there are new ideas and developments that may change textile cleaning in the future.

Technological innovation

As noted earlier, there has been active R&D on washing machines for many years, but radical technical innovation in washing machines has been rare. Instead there has been a stream of incremental innovations. For example, a major European Commission report as long ago as 1995 said that the energy-efficiency of washing machines could be increased by 25% using proven technologies such as dirt sensors, automatic detergent dosing, more efficient motors and longer, lower temperature washes (GEA, 1995).

Nevertheless, a few more radical innovations have managed to reach the market and may become established in the future. An example is detergent-free washing machines. An early example was a Sanyo water-electrolysing machine mentioned earlier. More recently the WasH2O machine, from Chinese multinational Haier, can clean with or without detergents. The WasH2O works by breaking water molecules into its OH- and H+ ions components. OH- acts as the cleaning agent by attracting and retaining stains while the H+ ions sterilise the clothes (Freshome, 2013).

A long anticipated, but more challenging, innovation is water-free or almost water-free washing, and some waterless devices have been developed. A concept design for a waterless washing machine, which cleaned using negative ions and compressed air, won the Electrolux Design Lab Award 2005. Sanyo launched its Aqua Loop machine which could dry clean and deodorise clothes and shoes using ozone generated from the air. The Aqua Loop can also wash with water, which can be recycled within the machine for subsequent washes. A machine has been developed by a UK company called Xeros that uses up to 80% less water, 50% less energy and half the normal amount of detergent to lift dirt from the wash-load, which is then removed and absorbed by 1.3 million reusable nylon polymer beads. The Xeros bead-cleaning system has been demonstrated in large commercial washers (Figure 2.14) with a domestic machine planned for launch in 2016 (Thompson, 2014).

IT is being applied in products like washing machines that goes beyond making the product more 'intelligent'. For example, trials are taking place with washing machines connected to smart meters or the Internet. The machine could then be loaded by the user, but turned on remotely by the energy supplier when the

FIGURE 2.14 Xeros large commercial washer which uses small amounts of water and detergent to lift dirt from the wash, which is then absorbed and removed by reusable polymer beads

grid is lightly loaded and electricity may be cheaper. Or the machine might be turned on at times that save most energy by providing information to the user when weather is forecast to be fine for drying the laundry or a domestic solar energy system will be generating electricity (Bourgeois *et al.*, 2014). Products from mainstream manufacturers are becoming smarter too. For example, Samsung's 2014 Ecobubble™ touchscreen machine has sensors to measure the size of the load and dirtiness of the wash in order to automatically select the programme and dispense the correct amounts of water and detergent. It can also be controlled remotely via a smartphone.

Other laundry innovations that could be developed, or adopted more widely, include: washing machines that reuse grey water from showers; reintroduction of hot and cold fill machines to utilise solar-heated water; microwave clothes driers; and stain- and dirt-resistant fabrics (Weaver *et al.*, 2000).

Sustainable laundry services and systems

Domestic washing machines are not the only way of doing the laundry. Some people have always sent their washing to a laundry, used a launderette or a shared apartment washing facility. Recently interest in alternative textile cleaning systems has emerged, based on the idea that what consumers want is clean laundry rather than better washing machines. It is argued that this service-based approach could reduce environmental impacts more than improving household washing machines, detergents and driers (Weaver *et al.*, 2000).

Innovations in ownership of laundry equipment

There are several possible changes in the ownership of equipment that might lead to a more sustainable laundry system:

- A system of leasing rather than direct purchase of washing machines to enable consumers to return their machines to manufacturers for recycling, or for upgrading to incorporate the latest technology. (The latter is similar to renting a television set, which after a period is returned for replacement with an upgraded machine.)
- A system in which consumers have a contract with a service company that provides a home washing machine and records the energy supply for operating the machine. The consumer pays per use of the machine and the service company maintains, upgrades and recycles the machines. (This has similarities to the 'pay-as-you-go' system for mobile phones.)
- Communal or commercial laundry facilities equipped with the most environmentally-efficient technologies. (Electrolux trialled such a system in the 1990s using long-life commercial machines which had automatic detergent dispensers and used recycled rinse water for pre-washing.)

The Netherlands government's Sustainable Technology Development programme sponsored research on sustainable clothes washing up to 2025 (van den Hoed, 1997; Weaver et al., 2000). The project examined whether a 95% improvement in 'eco-efficiency' could be achieved through technical innovation alone or through a shift from individually owned products to shared services. The shared services considered were a small neighbourhood wash centre and a large commercial laundry. The measures of eco-efficiency for all three scenarios were the amounts of primary energy, water and detergent consumed per kg of washed clothes.

The study indicated that improvements in household washing technology (e.g. low-temperature detergents and sensors to optimise machine loading) could produce a 60% reduction in energy consumption. The neighbourhood wash centre or large laundry could achieve a 55% reduction in energy consumption by using renewable energy or combined heat and power plus improved machines.

The study found that shared services only produced similar reductions in energy use to those at the household level, because some energy savings in laundries are offset by the need for higher wash temperatures, more artificial drying, and fuel for transport. Thus, about 60% reductions in energy per kg of clean clothes were achievable in all three scenarios, although the large laundry offered 90% reductions in water and detergent use by recycling water and detergent.

Beyond 2025, the study concluded, laundries could probably offer greater eco-efficiency improvements by use of new technologies. However, the study also recognised that there are major barriers in persuading people to shift to neighbourhood wash centres or laundries unless the cost and convenience was similar or better than home clothes washing. Consumer reluctance to change laundry behaviour was highlighted in a study which presented potential occupiers of a new UK housing development with the offer of a novel 'laundry to go' service. Households would not own washing machines, instead locking their dirty laundry in an outside bin from where it would be collected, laundered and delivered back to the bin (Figure 2.15). Although the householders liked other possible services like house cleaning offered in the package, they said they would prefer to do their own laundry because it was cheaper and more flexible (Dewberry et al., 2013).

New patterns of clothes ownership

If changing the technology and ownership of laundry equipment cannot achieve very major reductions in environmental impacts, perhaps changes to the way clothing is produced and consumed might? A major EU-funded project called The Sustainable Household (SusHouse) developed long-term future scenarios for household functions including Clothing Care (Green and Vergracht, 2002). Workshops were held in Italy, The Netherlands and Germany in which producers, consumers, technical experts and government representatives brainstormed ideas for sustainable futures for the production, consumption and care of clothing in the year 2050. The SusHouse project team then constructed scenarios intended to

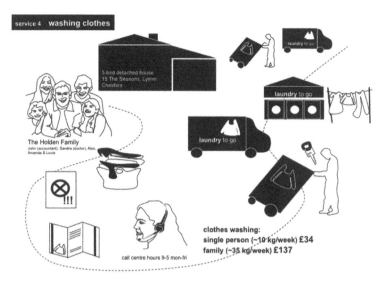

FIGURE 2.15 Concept design for a sustainable laundry service (from a draft of Dewberry *et al.*, 2013, Figure 3)

inspire today's designers to create future innovations. The scenarios that involved changes in clothes ownership included:

- Clothes leasing centres – some outer clothes are rented or loaned from service providers who also wash, clean, iron and repair them.
- 'My clothes, my eternal friends' – people own a limited wardrobe of high-quality, largely made-to-measure clothes, kept clean and repaired by a laundry service.
- Collective clothing care – high-quality clothes are shared and exchanged in a local clothing centre, where they may also be cleaned.

Of the three scenarios, consumer focus groups were most positive about the clothes-leasing system, which by sharing certain high-quality garments could reduce resource, water, energy and pesticide use and textile waste. They especially appreciated the possibility of professional clothes washing, repair and recycling.

However, it was difficult to communicate the idea of not only outsourcing laundering but also the ownership of clothes. Some consumers appreciated that they could have more variety with leased or rented clothes, but people who use clothes to present an image and express their personality had the most reservations. Consumer acceptance might therefore be increased by a mixed system: they could either use the total clothing care service, or they could just use the laundry service while still owning all or most of their clothes.

Some of these ideas have succeeded, at least at a micro scale. An example is the 'Keep and Share' knitwear business of a British designer–maker Amy Twigger Holroyd. Her garments are relatively expensive, handmade and classic knitwear made from carefully sourced materials, designed to be kept long-term, cleaned and repaired by Keep and Share if required and then handed down to family or friends (Chambers, 2012).

The conclusion of these futures studies is that innovative service-based systems of textile use and cleaning are difficult to introduce, given established habits and the cost and convenience of home-based machines. Also, because they still require physical equipment and energy, such innovations may only offer limited environmental benefits over conventional home-based systems, unless there is a breakthrough in textile cleaning technology that is only economic on a non-domestic scale.

The pattern of innovation

Many early electric washing machines were similar to manual machines, but driven by an electric motor. These early machines then developed via a huge variety of washing mechanisms and configurations before converging onto two dominant designs; the top-loading agitator and front-loading tumbler action machines. These dominant designs have subsequently been improved in numerous ways to improve their cleaning performance, reduce their environmental impacts and reduce their production costs. There are several new washing technologies and machines in small-scale production, but so far these innovations have not displaced the dominant designs.

References

Bourgeois, J., van der Linden, J., Kortuem, G. and Price, B. (2014) 'Conversations with my washing machine: interactive energy demand-shifting with self-generated energy', paper presented at the *Energy Research Conference*, Milton Keynes, The Open University, 3 April.

Bush, E., Damino, D., Josephy, B. and Granda, C. (2013) Heat pump tumble driers [Online], Available at www.topten.eu/uploads/File/EEDAL13_Eric_Bush_Heat_Pump_Driers.pdf (accessed December 2013).

Chambers, R. (2012) 'Amy Twigger Holroyd: Keep and Share', *Craft and Design*, November/December.

Consumers' Association (1958) 'Washing machines', *Which?* Summer, pp. 4–9.

Consumers' Association (1960) 'Large washing machines', *Which?* September, pp. 191–204.

Consumers' Association (1964a) 'Twin tub washing machines', *Which?* April, pp. 100–109.

Consumers' Association (1964b) 'Automatic washing machines', *Which?* November pp. 324–335.

Consumers' Association (1966) 'Automatic washing machines', *Which?* September pp. 276–286.

Consumers' Association (1968) 'Washing powders', *Which?* June pp. 169–74.

Consumers' Association (1969) 'Enzyme detergents', *Which?* September pp. 267–269.

Consumers' Association (1975) 'Washers and driers', *Which?* March, pp. 68–73.
Consumers' Association (1976) 'Washing machines', *Which?* February, pp. 42–46.
Consumers' Association (1979) 'Washing machines', *Which?* January, pp. 45–53.
Consumers' Association (1981) 'Twin tubs', *Which?* August, pp. 478–479.
Consumers' Association (1990) 'Which detergent?' *Which?* April, pp. 186-189.
Consumers' Association (1991) 'Which washing machine?', *Which?* January, pp. 48–57.
Consumers' Association (1994) 'Washing machines: best on test', *Which?* February, pp. 46–51. Note: *Which?* magazine changed from copyright Consumers' Association to copyright Which? Ltd. from April 1995.
DECC (2013) 'Energy consumption in the UK. Domestic data tables 2013' [Online] Available at https://www.gov.uk/government/publications/energy-consumption-in-the-uk (accessed December 2013).
Deni Greene Consulting (1992) *Life cycle analysis. Clothes washing machines*, Melbourne, Australian Consumers' Association.
Dewberry, E., Cook, M., Angus, A., Gottberg, A. and Longhurst, P. (2013) 'Critical reflections on designing product service systems', *The Design Journal*, vol. 16, no. 4, pp. 408–429.
Durrant, H.E., Hemming, C.R., Lenel, U.R. and Moody, G.C. (1991) 'Environmental labelling of washing machines. A pilot study for the DTI/DOE', Cambridge, UK, PA Consulting Group, August.
EPA (2013) 'Energy Star, US Environmental Protection Agency' [Online] Available at http://www.energystar.gov/index.cfm (accessed December 2013).
EST (2013) *At home with water*, London, The Energy Saving Trust, July.
Freshome Design and Architecture (2013) 'Detergentless Haier WasH20 Washing Machine' [Online] Available at http://freshome.com/2007/07/31/detergentless-haier-wash20-washing-machine/ (accessed December 2013).
GEA (1995) 'Washing Machines, Driers and Dishwashers Final Report', Group for Efficient Appliances, Copenhagen, Danish Energy Agency, June.
Green, K. and Vergracht, P. (2002) 'Towards sustainable households: a methodology for developing sustainable technological and social innovations', *Futures*, vol. 34, no. 5, June, pp. 381–400.
Johnson, A. (2013) *This boy*, London, Bantam.
Maxwell, L. (2009) 'Who Invented the Electric Washing Machine? An Example of how Patents are Misused by Historians' [Online] Available at http://www.oldewash.com/articles/Electric_Washer.pdf (accessed November 2013).
Maxwell, L. (2013) 'The magnificent machines of Monday', Powerpoint presentation [Online], Available at http://www.oldewash.com/ (accessed November 2013).
ONS (2011) 'Ownership of consumer durables increases into 2010' [Online], Office for National Statistics, UK, November. Available at http://www.ons.gov.uk/ons/dcp171780_245268.pdf (accessed December 2013).
Robertson, A. (1994) 'Smart consumer products with a pathfinder product development strategy', Proceedings Second European Conference on Smart Structures and Materials, Glasgow, p. 307, September.
Rosling, H. (2010) 'The magic washing machine', YouTube video of TED Talk, December. Available at http://www.youtube.com/watch?v=BZoKfap4g4w (accessed November 2013).
Roy, R. (2006) 'Products: new product development and sustainable design', T307 *Innovation* Block 3, Milton Keynes, The Open University.
Shove, E. (2003) *Comfort, cleanliness and convenience*, Oxford, Berg.

Shove, E. (2004) 'Sustainability, system innovation and the laundry', in Elzen, B., Geels, F.W. and Green, K. (eds) *System innovation and the transition to sustainability*, Cheltenham, Edward Elgar, pp. 76–94.

Spencer, J., Lilley, D. and Porter, S. (2012) 'The opportunities different cultural contexts create for sustainable design', Sustainable Design Research Group, Loughborough Design School, Loughborough, Loughborough University.

Thompson, S. (2014) 'Magic bead washing machines set to float', *The Times*, 22 February.

van den Hoed, R. (1997) 'A shift from products to services: an example of washing services', paper presented at Towards Sustainable Product Design, 2nd International Conference, Delft, The Netherlands Delft University of Technology, July.

Weaver, P., Jansen, L., van Grootveld, G., van Spiegel, E. and Vergracht, P. (2000) *Sustainable technology development*, Sheffield, Greenleaf.

Which? Ltd. (2002) 'Monotub titan', *Which?* January, p. 5.

Which? Ltd. (2005) 'Washing machines and detergents', *Which?* January, pp. 38–45.

Which? Ltd. (2011) 'Test lab: detergents', *Which?* January, pp. 44–45.

Which? Ltd. (2013a) 'Test lab: washing machines', *Which?* January, pp. 62–65.

Which? Ltd. (2013b) 'Test lab: washing powders', *Which?* July, pp. 56–57.

Which? Ltd. (2013c) 'Test lab: tumble driers', *Which?* November, pp. 64–67.

Which? Ltd. (2014) Samsung WF80F5E5U4W review [Online]. Available at http://www.which.co.uk/reviews/washing-machines/samsung--wf80f5e5u4w/review (accessed February 2015).

Wikipedia (2013) Washing machine [Online] Available at http://en.wikipedia.org/wiki/Washing_machine (accessed November 2013).

3
LAMPS AND LIGHTING

Illuminating the darkness when and where there is no natural light is one of humanity's essential needs. Since pre-history people have used flames to produce artificial light: first from fires, then from torches, candles and oil lamps (Figure 3.1a). In Europe and America from the late eighteenth century, gas lamps were introduced for home and street lighting (Figure 3.1b). At first gas lamps had a bare flame, but after the invention of the gas mantle in 1885, much brighter gas lamps were introduced in which the flame made a fabric impregnated with a rare earth mixture glow white hot (Encyclopaedia Britannica, 2013). However, by the early twentieth century the established system of gas lamps and coal gas production and distribution began to be challenged by the new system of electric lighting.

Outside Europe and America electric lighting also gradually spread to urban areas during the twentieth century, but in the twenty-first century there are still millions of people living in rural areas of developing countries who have no electricity or electric lighting. So they have to rely on oil lanterns or, in increasing numbers, have off-grid solar photovoltaic lighting systems or use solar battery lanterns.

Electric lamp technology

Electric lamps are deceptively simple, mass-produced consumer products. But their development has depended on numerous inventions, extensive trial-and-error experimentation, the application of advanced scientific knowledge and organised R&D programmes as well as much materials and manufacturing process innovation. The use of electric lamps also depended on the development and installation of whole new systems of electricity generation and distribution.

FIGURE 3.1 Pre-electric lamps
(a) nineteenth-century oil lamp
(b) Ancient to nineteenth-century oil and gas lamps

The incandescent lamp

Modern lamps and lighting began with the invention of the incandescent electric lamp. An incandescent lamp is one in which a filament gives off light when heated by an electric current. However, incandescent lamps were not the first lamps to use electricity. Arc lamps, in which light was produced by an electric arc passing between carbon electrodes, were demonstrated in the early nineteenth century. But as they required large amounts of power their practical application had to wait until the development of electrical generators in the 1860s and 1870s. Arc lamps produced a very bright light only suited to large-scale uses such as lighting streets, factories and railway stations. Also, the carbon rods were consumed as the lamp burned and so had to be replaced frequently.

Carbon filament electric lamps

The American inventor–entrepreneur, Thomas Edison, and others realised that arc lamps were unsatisfactory, especially for illuminating small spaces. Edison thus saw the need to 'subdivide the electric light' and produce a small electric lamp that could be used in the home, in shops, etc. (Smithsonian Institution, 2008). Edison's understanding of the potential market and what was required in an electric lamp for domestic and small-scale commercial lighting shows the need for an inventor or designer to 'frame' a problem correctly to produce a successful innovation.

In the decades before Edison patented a carbon-filament incandescent lamp in 1880 (Figure 3.2), many scientists and inventors had attempted to produce a practical incandescent lamp. Notable among them was an Englishman, Sir Joseph Swan. In 1850 Swan produced an incandescent lamp with filaments made of carbonised paper; later he used filaments made from cotton thread treated with sulphuric acid. After 1875 the filaments were enclosed in glass bulbs from which the air had been evacuated.

The development of the incandescent lamp was the result of simultaneous work by Swan in England and Edison in the United States. Edison began studying the problem in 1877 and within 18 months he and his team of scientists, engineers and

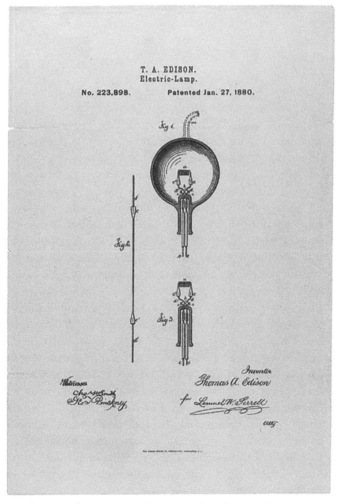

FIGURE 3.2 Thomas Edison's patent drawing and application for an improvement in electric lamps, patented 27 January 1880

technicians at his Menlo Park laboratory had experimented with more than 1,200 filament materials. Success, however, relied on the invention and development of an effective vacuum pump by a German, Hermann Sprengel. This pump was the key tool which made it possible, in 1879, to sufficiently exhaust the air from a light bulb so that a carbon filament lasted long enough to be useful. In October 1879, Edison switched on a lamp containing a carbonised thread filament, which lasted 14.5 hours. Later he discovered that filaments of carbonised bamboo would last several hundred hours and adopted this as his preferred material. Both Swan's and Edison's lamps consisted of a carbon filament in an evacuated glass bulb with the ends of the filament's connecting wires brought out through a sealed cap to the electricity supply. When the supply was connected, the filament glowed and, because of the vacuum, did not oxidise away quickly as it would have done in air. Edison and Swan both applied for patents for their incandescent lamps, and the ensuing patent dispute between the two men was resolved by the formation of a joint company in 1883 (Edison Tech Center, 2013).

The electric lamp was, however, of little use without a proper electricity supply. Consequently, Edison's team set about designing and installing an entire electricity generation and distribution system. This is what distinguished Edison as an innovator from Swan. Swan used a separate generator for each property with electric lamps. Edison realised that for electric lighting to enter widespread use, a system of supplying electricity to a whole area was needed. Edison has therefore received the main credit for inventing the electric lamp, because at the same time as developing incandescent lamps he also developed electric generators, power lines and the multiple circuit equipment needed for a practical lighting system. Edison demonstrated his system to financiers and industrialists and to politicians to get approval for city installations. From the original demonstration in the grounds of his Menlo Park Laboratory in late 1879, Edison's system of incandescent electric lighting began to diffuse into use. The first commercial installation was made in May 1880 on the steamship Columbia, and in 1881 a New York City factory was lit with Edison's electric lamps. In 1882 Edison switched on the power from his first electric power generation plant in Manhattan, illuminating 800 carbon filament bulbs in the area, thus beginning to fulfil his vision for urban electric lighting (Open University, 2014).

Edison's and Swan's carbon-filament lamps were highly inefficient but avoided the dirt and fire hazards of gas lamps, and so electric lamps started to gain acceptance by the early twentieth century. The difficulty and cost of installing the electricity distribution system, however, slowed their adoption. Hence the ability to use electricity for non-lighting tasks became an important incentive to install electricity in the home. As the electricity network grew manufacturers developed a range of domestic electric appliances, including washing machines (Chapter 2), refrigerators and electric heaters (Smithsonian Institution, 2008). The introduction of the electric lamp and the development and adoption of other electrical appliances therefore reinforced each other.

Tungsten filament electric lamps

The most important improvement to the incandescent lamp was the use of metallic filaments, first tantalum and then tungsten. Tungsten is especially suitable for filaments because of its high melting point. This means that tungsten filament lamps can operate at higher temperatures and therefore emit more light for the same electrical input than is possible with carbon filaments. The first tungsten filament lamps, introduced in the United States in 1907, made use of pressed tungsten. By 1910 a more economical process for producing drawn tungsten filaments had been developed by General Electric. In the early twentieth century tungsten filament lamps began to be made in large volumes and gradually eliminated the surviving competition from gas lighting (Figure 3.3a).

The early tungsten lamps, however, suffered from filament evaporation, causing blackening of the bulb, and thinning until the filament broke. Around 1913 it was found that a small amount of argon or nitrogen gas in the bulb reduced evaporation and enabled the filament to run at a higher temperature, giving a whiter light, higher efficiency and a longer life. Other improvements followed, including the development of the coiled filaments that further increased efficiency (Smithsonian Institution, 2008). Beginning in 1925, some bulbs were frosted on the inside to provide a diffused light. The double-coiled filament was introduced around 1930 and increased efficiency a bit further because the more compact form reduced heat losses (Edison Tech Center, 2013). Single coil and 'coiled-coil' tungsten filament lamps became the standard 'general lighting service' (GLS) incandescent lamps which dominated domestic lighting for the next 70 years (Figure 3.3b).

The importance of manufacturing

Inventing a product often calls for the development of new manufacturing processes, which in turn has often led to design changes in the product. Early incandescent

FIGURE 3.3a Early twentieth-century tungsten filament incandescent light bulbs

FIGURE 3.3b Early twenty-first-century clear and frosted tungsten filament incandescent (GLS) light bulbs with bayonet and screw fittings

52 Lamps and lighting

lamps were assembled by hand and were consequently too expensive for most consumers. Edison recognised this problem and produced several manufacturing process patents aimed at mass-producing electric lamps (Smithsonian Institution, 2008).

Eventually, automatic machinery was developed to make standard GLS lamps that were designed for automated manufacture and their cost fell dramatically. In manufacturing the bulb, a continuous ribbon of glass is passed along a conveyor belt, heated in a furnace, and then blown into moulds by air nozzles through holes in the conveyor belt. After the bulbs have cooled, they are cut off the ribbon machine. The filament supports are assembled on a glass stem, which is then fused to the bulb. The air is pumped out of the bulb through a tube, and flame sealed. The bulb is then inserted into the lamp base, and the whole assembly tested. A typical production system produces 50,000 lamps per hour (Advameg, 2014).

The tungsten halogen lamp

After the Second World War the search for higher efficiency and durability of electric lamps led to the development of the tungsten halogen light bulb. This is an incandescent lamp that has a small amount of a halogen such as iodine or bromine gas added. The halogen gas and the tungsten filament create a chemical reaction that re-deposits evaporated tungsten onto the filament, increasing the lamp's durability and efficiency. To allow this reaction the bulb has to able to withstand high temperatures, so quartz or aluminosilicate glass is used for halogen lamps. The first commercial tungsten halogen lamps were launched by General Electric in 1959. To achieve their high operating temperatures, tungsten halogen lamps are much smaller than ordinary GLS light bulbs. The most common designs are standard GLS lamps enclosing a halogen capsule and halogen spot lamps extensively used in kitchens and bathrooms (Figure 3.4). Xenon-filled and infrared-coated halogen lamps offering better efficiencies than early tungsten halogen lamps were subsequently developed and introduced.

Dominant design

With tungsten halogen lamps and incremental improvements to the ordinary GLS tungsten light bulb, the incandescent lamp became the dominant form of domestic and commercial electric lighting until the end of the twentieth century.

This dominance of the ordinary GLS tungsten lamp for domestic lighting is reflected in *Which?* reports. Before 1980 the magazine concentrated on reviewing GLS lamps in its reports that compared the price, life and running costs of standard 1,000-hour and double life single coil and coiled-coil bulbs (Consumers' Association, 1975, 1978, 1979). Although cheap to buy (16p to 25p for a 60W bulb in 1978), a GLS tungsten lamp only converts about 5% of the energy it uses into visible light (an efficiency of about 15 lumens per watt), the rest being converted into heat. Tungsten halogen lamps offer 20% to 45% greater efficiency

FIGURE 3.4 Modern tungsten halogen incandescent light bulbs
(left) GLS light bulb with tungsten halogen capsule
(right) Tungsten halogen reflector spot lamp with GU10 connector

and so were permitted as direct replacements for GLS tungsten lamps as sales of the latter were phased out in some countries from 2005 and from the EU between 2009 and 2012.

Fluorescent lamps

During most of the twentieth century the incandescent lamp was the dominant technology. But during and after the Second World War a new lighting technology, the fluorescent lamp, offered an alternative for domestic, commercial and industrial lighting.

To a greater extent than the incandescent lamp, the fluorescent lamp depended on scientific discoveries and technical inventions. A significant landmark occurred in 1855 when a German named Geissler partially evacuated the air from a glass tube and passed an electrical current through the tube, creating a strong green glow at one end of its walls. The Geissler tube, in which gas at low pressure glows when subjected to an electrical voltage, demonstrated the principle of the electric discharge lamp. In 1895 a former employee of Edison, Daniel Moore, demonstrated discharge lamps that made carbon dioxide gas in glass tubes emit white light. Although Moore's lamp was complex and expensive, and required high voltages, it was more efficient and produced a more natural light than incandescent lamps, and so in the early twentieth century Moore's lighting system was installed in some shops and offices. Around the same time, another American, Peter Cooper Hewitt, patented the mercury-vapour discharge lamp, another important landmark in the development of the modern fluorescent lamp. Hewitt's lamp glowed when an electric current was passed through low-pressure mercury vapour and, unlike Moore's lamps, operated at low voltages. The mercury-vapour lamp was more efficient than the incandescent lamps of the time, but the blue-green light it produced limited its use.

54 Lamps and lighting

The next important innovation in discharge lighting used tubes of neon gas, which glowed bright red. The development of the neon light also helped provide the final technical step for the fluorescent lamp when in 1926 Jacques Risler patented fluorescent coatings on the inside of neon light tubes. The main use of neon lamps, however, was for advertising signs, not general lighting (Edison Tech Center, 2013).

Thus after decades of research and invention, the technologies of mercury vapour discharge lamps and fluorescent coatings were available. With knowledge of these key technologies, and stimulated by a report of successful experiments with fluorescent lighting in Britain, a team of General Electric engineers in America produced a prototype fluorescent lamp in 1934. In 1936 GE demonstrated its 'Lumiline' fluorescent lamp – which produced a bright green light – at a banquet to celebrate the centenary of the US patent office. In 1938 GE launched the first 'Mazda' fluorescent lamps in various colours including white, which were rapidly developed until the standard 4-foot 40W white fluorescent tube was launched in 1939 (DeLair, 2013).

The fluorescent lamp consists of a glass tube filled with a mixture of argon and mercury vapour. When current flows through the gas between the electrodes, the gas is ionised and emits ultraviolet radiation. The inside of the tube is coated with phosphors (substances that absorb ultraviolet radiation and fluoresce), producing visible light. Because a fluorescent lamp does not provide light through the heating of a filament, it is four to six times more energy efficient than an incandescent lamp. However, a high voltage is needed initially in order to ionize the gas. This voltage is supplied by a device called a ballast, which also maintains a lower operating voltage after the gas is ionised (Edison Tech Center, 2013).

Social shaping of fluorescent lighting

General Electric's development of fluorescent lighting is one of the case studies in a book on the theory of the 'social construction of technology' or SCOT (Bijker, 1995). As was briefly discussed in Chapter 1 for the case of the safety bicycle, this theory argues that innovations do not evolve through the steady linear progress of science and technology. Instead, technological innovation is 'shaped' by the social, economic and political interactions of relevant social groups (RSGs or 'actors') in society. In support of the theory Bijker shows that when General Electric (GE) tried to introduce the fluorescent lamp, originally as a source of specialist colour lighting and then as a source of high-efficiency general lighting, the company came up against opposition from American electric utilities who feared that the lamp's high efficiency would harm their electricity sales. However, after fluorescent lamps had been demonstrated in 1939 at major American public exhibitions, consumers and lighting engineers were so eager to obtain this lighting innovation that the utilities were forced to accept some form of fluorescent lamp. After intense disputes between the manufacturers and utilities they came to an agreement that fluorescent lamp development would focus on providing high-intensity rather than high-efficiency lighting so as not to harm electricity sales. This agreement led in turn to the US government suing GE and the utilities for forming a cartel, which GE successfully countered by arguing that the court case would harm the War effort. Bijker concludes that the types of fluorescent lamp first developed and introduced on the market were the outcome, not of straightforward attempts by scientists and engineers to develop a better lamp or even of manufacturers to seek profit through innovation, but of bitter power struggles between the relevant social groups, namely manufacturers, utilities and government, influenced by engineers and consumers.

Adoption of fluorescent lamps

Despite the commercial and political arguments, the need for lighting of factories during the Second World War justified the production of energy-efficient as well as high-intensity fluorescent lamps. But it would not be until the 1950s that efficient fluorescent tubes would be diffused into widespread industrial and commercial use, overtaking incandescent lamps as the main lighting source, and also start being used for domestic lighting, mainly in kitchens, bathrooms and garages.

An indication of the lack of interest in domestic applications was that until 1980, *Which?* only briefly mentioned fluorescent lamps in its lighting reports. For example, a 1975 report noted that fluorescent tubes would save costs over their life-time, but required special fittings and produced lighting colours that consumers might not like (Consumers' Association, 1975). In 1979 *Which?* again mentioned the long life, high efficiency and drawbacks of fluorescent tubes, while announcing the new 30W Electronic Halarc metal halide lamp launched by GE in America, designed as a possible replacement for ordinary GLS lamps (Consumers' Association, 1979). Metal halide lamps, however, never caught on in competition with the new compact fluorescent lamps being developed at the same time. By 1980 *Which?* finally recognised that straight or circular fluorescent lamps (Figure 3.5) had many uses around the home, especially in kitchens and garages and as concealed and display lighting in living areas at four times the efficiency of GLS lamps (Consumers' Association, 1980).

FIGURE 3.5 Circular fluorescent lamps – decorative, but much less common than straight fluorescent tubes

Compact fluorescent lamps

The energy crisis that followed the 1973 rise in oil prices spurred research and innovation to produce more energy-efficient lamps. As noted previously, fluorescent lamps are much more efficient than incandescent ones. Thus in the 1970s, many engineers proposed designs for compact fluorescent lamps (CFLs) that could replace the ubiquitous GLS incandescent light bulb. Although these designs often worked in the laboratory, most were considered too expensive to mass-produce and so were not put into production. For example, in 1976 Edward Hammer, an engineer working for General Electric in America, had the idea of bending a long, thin fluorescent tube into a spiral shape. This not only allowed for the long electrical arc necessary for the lamp's operation, but also produced light similar to that of a frosted incandescent lamp. Existing lamp machinery, however, had difficulty making the spiral, and GE felt that the $2 million investment required for new production equipment was too high, so they shelved the design and focused on development of the metal halide lamp, mentioned earlier (Smithsonian Institution, 2008).

However, spiral fluorescent lamps began to appear on the market as other manufacturers decided to see if a competitive design could be produced. European efforts, led by the Dutch multinational Philips, were directed at producing a commercially viable CFL. By 1976 Philips had produced a prototype, the SL1000 (Self-ballasted Luminescent 1000 lumen) lamp, which then took five years to develop into a commercial product. A scientific advance that made this lamp feasible was the development of rare earth phosphors by Philips researchers which slowed the rapid deterioration of light output that affected small diameter tubes with traditional phosphors. It also required substantial investments in new production equipment to enable the lamps to be made reasonably economically.

In 1980 Philips launched the first mass-produced CFL, the Philips SL18, designed as an energy-efficient alternative to the incandescent lamp and which would fit into a standard lamp holder (Figure 3.6a). The lamp contained a small-diameter fluorescent tube folded into two U-shapes with an Edison screw or bayonet cap fitting. A magnetic ballast to regulate the electric current was built into the base of the lamp and the tubes were enclosed in a thick glass jar, making the lamp about the size and weight of a can of baked beans (Museum of Electric Lamp Technology, 2013). *Which?* tested a sample of the new 18-watt lamps in 1981, confirming that they produced more light than a 60-watt GLS lamp, saving over 70% electricity, and on the way to meeting their 5,000 hours' life. Although the SL lamp is a significant lighting milestone, it was slow to penetrate the market. A major factor, because of the large investments in research and manufacturing equipment, was its cost, about £8 to £9 when first introduced to the UK market, while a GLS lamp only cost from 25p. However, even these early CFLs would have saved consumers money. *Which?* calculated that over 5,000 hours a Philips SL18 would save about £3 in purchase and electricity costs compared to five 1,000-hour 60W GLS lamps (Consumers' Association, 1981, 1985). However, this small long-term saving was not a sufficient incentive for most consumers to adopt CFLs. In addition, the weight of the early CFLs made table lamps unstable and their size meant that they would not fit small shades. Their slow warm-up also made the new lamps unacceptable to many consumers.

In order to promote energy saving and to kick-start the market, governments and electric utilities offered CFLs to householders free or at reduced prices, which helped to diffuse them and encouraged other companies to enter the market (Stokes *et al.*, 2006). By 1993 *Which?* had conducted tests of the various CFLs that had been introduced. These included stick and flat D-shape designs, with integral or separate magnetic or electronic ballasts, and standard light bulb shapes. Using just 7W to 23W of power, they produced light outputs equivalent to 40W to 100W incandescents. And despite costing £6 to £18, *Which?* calculated that after 7,500 hours of use, worthwhile savings of £10 to £30 could be made by replacing several incandescents with an equivalent CFL (Consumers' Association, 1993). By 1999 *Which?* reported on smaller and lighter versions of CFLs with integral electronic controls in various designs that came on instantly and reached full brightness faster. Prices and payback periods had also come down (Which? Ltd., 1999).

But although the market for CFLs grew ten-fold from 1990 to 2000 with about 1.3 billion in use, only about 5% of lights in the EU were CFLs, the rest being incandescent lamps (Wilhite, 2001). By the start of the withdrawal of incandescents from the EU market in 2009 *Which?* viewed the CFL, now reduced in price, available in many shapes and fittings and lasting at least 8,000 hours (Figure 3.6b) – as the main type of electric lamp for domestic use (Which? Ltd., 2009) with LED lamps for general lighting still waiting to be fully introduced.

FIGURE 3.6 Compact fluorescent lamps
(a) Philips SL18, the first mass-produced CFL, 1980
(b) Modern CFL lamps with screw and bayonet connectors and (right) CFL spot lamp with GU10 connector

LED lamps

A light-emitting diode, or LED, is a device that operates on completely different principles from incandescent and fluorescent lamps. It directly converts electricity into light by electroluminescence of a semiconductor material. Electroluminescence is the phenomenon of a material emitting light when subjected to an electric current or electric field. An LED comprises a small semiconductor crystal with reflectors and other parts to make the light brighter and focused onto a single point. One or many LEDs combined with reflectors, housing and other components create a complete LED lamp (also called an LED module or package).

The phenomenon of electroluminescence was discovered in 1907, but the first modern LEDs, which produced infrared light, were developed at Texas Instruments in 1961 and first used by IBM to replace incandescent lamps in computer-punched card readers. The first LED to produce visible light was a red LED developed at General Electric in 1962. This was followed by yellow, violet, and by 1979 blue, LEDs each based on different semiconductor materials. However, it wasn't until the 1990s that LEDs could be produced at a cost able to justify commercial production. LEDs began to be used in many applications, such as coloured electronic displays and bluish decorative lighting. To produce a white LED lamp a phosphor coating was used to convert blue or ultraviolet LED light, or the light from different colour LEDs was mixed. However, white LED lamps with sufficient output for general lighting (Figure 3.7) did not become an affordable product until the early twenty-first century (Edison Tech Center, 2013).

Which? did not mention LEDs until 2010, when it reported on two LED lamps with relatively low outputs of 200–300 lumens, less than an 8W CFL or equivalent to a 25W incandescent (Which? Ltd., 2010). By mid-2013, after ordinary GLS incandescents had been removed from sale in the EU, *Which?* had tested eight LED lamps, indicating that they were becoming a mainstream product. They were priced at £7 to £35 and available in outputs up to 650 lumens, equivalent to a 40W incandescent. By late 2013 LED lamps with over 800 lumens output, equivalent to a 60W incandescent, and with higher efficiencies were tested, showing how fast the technology was developing (Which? Ltd., 2013a; 2013b). Not all LEDs, however, lived up to their claimed life of 15,000 hours or more. *Which?* reported on tests that showed that nearly 30% of LED lamps failed before 10,000 hours and some did not reach the EU standard of a 6,000 hours' minimum life (Which? Ltd., 2014a). Such shorter lives seriously undermine the economics of LED lamps because their two to three times higher cost is only paid off if the lamp reaches a lifetime of 25,000–30,000 hours. A typical lifetime of 15,000 hours requires a considerably lower retail price to make LED lamps more cost-effective than CFLs. Even so, it is anticipated that solid state lighting, principally LEDs with their advantages of high and increasing energy efficiency, instant brightness, no mercury content and compactness will be the main source of domestic and commercial lighting in the future (US DOE, 2013a).

FIGURE 3.7 LED lamps with standard screw and bayonet connectors and (right) a 16-LED spot lamp (a direct replacement for a GU10 halogen spot lamp)

Lamp and lighting design

As shown in this brief account of lighting technology, there are many designs of electric lamp. But their design has largely been determined by the requirements of their technologies, methods of manufacture and compatibility with electricity supplies and light fittings. The basic form of the incandescent light bulb was established in the late nineteenth century and remained essentially unchanged until its steady withdrawal in the early twenty-first century. Other forms also reflected the requirements of their technologies. For example, LEDs emit light in a single direction so LED lamps have to incorporate reflectors and lenses to provide a useful spread of light. They also need heat sinks to avoid overheating, which would change their colour and reduce their life, and be designed to last for many years.

In addition to the requirements of technology, many incandescent halogen, CFL and LED lamps had to be compatible with established screw and bayonet connectors and fit existing lamp shades and light fittings. However, new types of connector were also developed for the new types of lamp, such as the GU10 bayonet connector for halogen spot lamps, which LED equivalents had also to fit (Figure 3.7 right). In addition, if they were to fully replace incandescent lamps, equivalent forms of CFL and LED had to be produced, including candle and mini globe designs.

However, new lamp technologies do not necessarily have to simply replace existing ones. In particular the low power, cool operation and the variety of shapes and sizes of CFLs and LEDs have given designers freedom to create novel designs of complete lamps and luminaires. Examples include CFL and LED desk lamps and LED strips to illuminate cupboards (Figures 3.8a and 3.8b) as well as solar-powered LED lamps for rural areas of developing countries, mentioned at the beginning of this chapter.

FIGURE 3.8a Stick CFL desk lamp

FIGURE 3.8b LED wardrobe light that comes on automatically when the door is opened

Environmental impacts and regulation

Domestic lighting is a relatively small, but nevertheless important, user of energy and source of greenhouse gas emissions and other environmental impacts. Lighting accounted for 13% of EU and US domestic electricity use in 2006 and 2011 respectively (EC, 2009; US Energy Information Administration, 2013). In the UK lighting accounted for 16.3% of domestic electricity use in 2012, although only 3% of total domestic energy use, down from 19% in 1990 (DECC, 2013a).

Lamp efficiency

Electric lamp efficiency is measured in lumens per watt. Very useful and detailed guidance for choosing energy-efficient lighting is provided by the PremiumLight Project carried out by the Austrian Energy Agency in partnership with 12 other EU organisations (PremiumLight, 2014). The guidance covers not only the energy efficiency of incandescent, halogen, fluorescent, CFL and LED lamps but their advantages and disadvantages in the colour (warmth) of light they produce and how accurately they reproduce colours, plus their lifetime costs (Tables 3.1 and 3.2).

PremiumLight reveals that different makes and designs of lamp have a wide range of energy efficiencies. The CFLs recommended for domestic use range in efficiency from 44 to 70 lumens/watt and LEDs from 46 to 85 lumens/watt, so careful consumer choice is important for maximum benefit. *Which?* also conducts technical tests of electric lamps and found some of its 2014 Best Buys, all of which were LEDs, now offer efficiencies higher than those in PremiumLight's tables, albeit at a price. The most efficient lamp tested (a 10W LED) provided 1,008 lumens (equivalent to a 70W incandescent, but using 86% less electricity) at an efficiency of 100 lumens/watt. It has a claimed 15-year life, but cost £15 (Which? Ltd., 2014b). *Which?* also recommended replacing halogen spot lamps with LED

TABLE 3.1 Advantages and disadvantages of different energy-efficient lamps

LED lamps	+ 80–90% reduced energy consumption + 10–30 times longer lifetime + Significant cost savings over lamp's lifetime - Expensive purchase price - Light distribution less 'wide'
Compact fluorescent lamps and tubes	+ 80% reduced energy consumption + 10–15 times longer lifetime + Comparably low purchase price + Significant cost savings over lamp's lifetime - Significant warm-up time - Contains mercury
Halogen lamps	+ Low purchasing price + Very good colour rendering + Brilliant light - Low efficiency and lifetime - High operating costs

TABLE 3.2 Efficiency, life and lifetime cost of typical electric lamps

Criteria	Incandescent	Halogen	CFL	LED
Luminous flux (lm)	660	700	740	810
Power (W)	60	46	14	12
Efficiency (lm/W)	11	15	52	67
Efficiency class	D	C	A	A
Lifetime (hrs)	1,000	2,000	10,000	30,000
Purchase price (€) 10 years [1]	10	20	9	10
Energy cost (€) 10 years [1]	72	55	17	14
Total Cost (€) 10 years [1]	82	75	26	24

[1] Assumption: operational time 1000hr/year

equivalents, such as a 7W LED costing £11 to replace a 50W halogen (Which? Ltd., 2014c). Lower priced LEDs from £4–£6 were also available.

The most efficient CFLs *Which?* tested were also improving in performance. For example, a Philips 12W spiral CFL provided 765 lumens (similar to a 60W incandescent lamp) at 64 lumens/watt, with a claimed 10-year life. At £3.50 it cost less than most LEDs (Which? Ltd., 2014b). However, the efficiency of LEDs is expected to continue to improve significantly and their costs to fall, while that of CFLs is levelling out (US DOE, 2013b).

Life cycle impacts

Life cycle assessment (LCA) studies of the environmental impacts of different types of lamp show clearly that their main impacts are overwhelmingly due to the energy consumed during the *use* phase of the lamp's life cycle. An early LCA study by

Muis et al. (1990) showed that per unit of light a fluorescent tube produces about a sixth, a CFL about a quarter and a tungsten halogen incandescent lamp about half of the amounts of air pollution and solid waste of an GLS incandescent lamp. It showed that even the mercury emissions of using GLS incandescent lamps are greater than those of making, using and disposing of fluorescent lamps, which contain small amounts of mercury. This is because burning coal, which includes trace amounts of mercury, to generate electricity releases mercury into the air and GLS incandescents use much more electricity.

Since then a number of LCA studies have been conducted that compare different types of lamp. For example, a report by the US Department of Energy (US DOE, 2013a) drew on existing LCA studies to compare the total life-cycle impacts of providing the same amount of illumination (20 million lumen-hours) from 1,000-hour life 60W GLS and halogen incandescents replaced many times, 8,500-hour life 15W CFLs replaced three times and one 25,000-hour 12.5W LED lamp. The review showed that the average life-cycle energy use of the LED lamp and CFLs was similar at about a quarter to a third of the energy use of the incandescent lamps. It found that the use phase of all four types of lamp accounted, on average, for over 90% of total life-cycle energy use followed by the energy for manufacturing and transport. Most of the uncertainty in the life-cycle energy use of LED lamps concerned their manufacture. Different sources estimated this at anywhere from 0.1% to 27% of the total. But if LED lamps meet their 2015 targets for improved efficiency and durability, their life-cycle energy use is expected to decrease by approximately 50% (Figure 3.9).

The same US DOE report also gave the results of a new LCA that compared the three types of lamp on 15 environmental impacts from climate change and water pollution to hazardous landfill waste. Unsurprisingly, incandescent lamps were found to be the most environmentally harmful across all impacts the DOE examined. The CFL was found to be slightly more harmful than today's LED

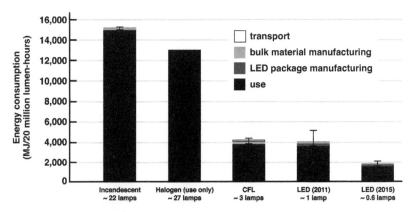

FIGURE 3.9 Life cycle energy use of GLS and halogen incandescent, CFL and LED lamps (Adapted from: US DOE, 2013a)

lamp on all impact measures except hazardous waste, because of the materials-processing impacts of the LED lamp's aluminium heat sink. As the performance of LED lamps continues to improve, aluminium heat sinks are expected to get smaller and recycling efforts could reduce their impact even further. The light source that performed the best was the LED lamp projected for 2017, whose total impacts are expected to be about 50% lower than the 2012 LED lamp and 70% lower than the CFL (US DOE, 2013b).

Environmental legislation

As mentioned earlier, sales of incandescent lamps have been phased out in many countries, starting with Australia in 2005. From 2009 electric lamps sold in the EU all had to meet minimum energy-efficiency standards – an 'A-class' as defined in the EU Energy Label for frosted, and a 'C-class' for clear, household light bulbs (EC, 2009). This meant that all frosted incandescent lamps had to be withdrawn from the market from 2009 and by 2012 sales of all clear bulbs had to be stopped. Improved incandescent lamps – halogen lamps with a xenon gas filling offering 25% better efficiency (C-class) than the best incandescents, and low-voltage halogen lamps with a separate or built-in transformer and an infrared coating offering 45% better efficiency (B-class) – were still permitted, as were CFL and LED lamps offering 65–80% better efficiency (A or B-class). From 2013 A+ and A++ classes were added to the Energy Label to cover the most efficient LED lamps and other lighting technologies (Figure 3.10). In the US energy-efficiency standards for lighting were introduced that, likewise, phased out tungsten incandescents from 2012–14 in favour of halogen lamps, CFLs and LEDs.

FIGURE 3.10 EU Energy Labels for halogen incandescent (C-class), CFL (A-class) and LED (A+class) light bulbs

Consumer adoption of energy-saving lamps

UK Energy statistics reveal a major twenty-first-century shift in the types of electric lamp used in British households. In 2000 GLS incandescent light bulbs accounted for 78% of domestic lighting electricity use; by 2012 the figure was down to 12%. Only slightly more efficient tungsten halogen lamps accounted for 7% of UK domestic electricity use in 2000 but had increased to 50% by 2012, mainly due to a huge increase in installations of 35W and 50W halogen spot lamps. CFLs used 4% of UK domestic electricity in 2000, rising to 28% by 2012, while fluorescent tubes used 12% of electricity in 2000 down to 9% by 2012. Virtually no LEDs were used in British homes in 2000 and by 2012 they accounted for less than 1% of domestic electricity use. Total UK domestic lighting electricity use fell by 12% between 2000 and 2012 (DECC, 2013b). However, the energy savings due to more efficient lamps were partly offset by increases in the number of lamps per household and number of households.

Lamps are therefore one of the clearest examples of how technical innovation reinforced by environmental legislation has reduced energy use and other environmental impacts of a product. There is great potential for further improvements with more widespread adoption of CFLs and LEDs; but the UK statistics show that there is still a long way to go to replace halogen incandescent lamps, especially multiple halogen spot lamps in kitchens, etc., with LED equivalents (such as those in Figure 3.7).

Social influences and impacts

In the late nineteenth and twentieth centuries the introduction of electric lighting led to dramatic social changes. The first was people's much greater control over interior lighting and hence the ability to continue their activities after dark and move away from old biological rhythms of waking and sleeping. The second was the infrastructure that brought electricity into homes and made it economical to introduce a wide variety of electrical appliances. A third consequence was to facilitate night working and leisure activities. Electric lighting also increased architectural possibilities as daylight became only a supplementary source of light, permitting the construction of much larger and taller buildings. Along with the development of electric lamps, designers and manufacturers created an increasing variety of light fittings (luminaires) that provided control over the quality and direction of light and offered fashionable designs. Lamps and lighting thus became an essential element of interior design (Smithsonian Institution, 2008).

One of the main drivers for innovation in lighting has been the technology 'push' for more energy-efficient lighting, stimulated by energy policies and energy-efficiency standards and legislation. However, the take-up of energy-saving lamps has been affected by a number of social factors and preferences. For example, consumer resistance to adopting early models of CFL due to their dimness, slow warm-up, light quality, size and incompatibility with existing light fittings has already been noted. These problems, which made CFLs the butt of many jokes, have largely been overcome, but the prejudice against them still persists among some consumers.

A study by Stokes et al. (2006) discusses some of the social factors that influence UK domestic lighting demand. One factor, they say, is an increase in the number of lamps per household, as UK consumers changed from a single ceiling light to several lights around the room. Another factor is the growing importance of lighting in the home. The authors reported interviews with consumers which revealed that 'flexibility', a 'better atmosphere' and a desire 'to create different moods' were factors in consumer decisions about lighting. Style was quoted by many respondents as being the main driver when selecting light fittings and lamps and several mentioned style programmes on television and articles in magazines as being highly influential in their lighting decisions. Style was usually the reason for choosing multiple halogen spot lamps for kitchens and bathrooms, even though some interviewees were aware that this choice wastes energy. One respondent, for example, said 'I was fully aware … that halogen lights are not … ecologically sound, it was a specific style choice.' Some interviewees mentioned dimmer switches – which are incompatible with many CFLs – to change lighting levels, while older people mentioned the need for brighter lighting. The authors conclude that 'respondents … were far more swayed by style … than by environmental issues.' This has meant that the potential savings from energy-saving lamps have often been less than expected, as consumers prioritise style and appearance and preference for the light provided by halogen incandescent lamps over the environment.

A related issue is the 'rebound effect', in which consumers choose to install more lights and/or leave them switched on longer if they choose CFLs or LEDs that are cheaper to run and last longer. This is an issue that is recognised in official calculations on the energy savings expected from energy-efficient lighting. For example, the assessment of the 2009 EC regulation phasing out GLS incandescent lamps assumed that a 15% increase would take place in the number of domestic lamps installed between 2007 and 2020. Despite the rebound effect, the assessment forecast that the electricity used for EU domestic lighting should decrease by 14% over the same period.

Future developments

Technological innovation

The long history of lighting innovation continues and there are lighting technologies other than those discussed in this chapter, such as metal halide and sulphur lamps. One of the most promising new technologies, however, is organic light emitting diodes (OLEDs) formed into flat panels.

Organic light emitting diodes

There are several current research programmes on OLEDs. Most existing applications of OLEDs are for mobile phone and TV screens (see Chapter 4). OLEDs may not be economical for general lighting for some time as currently available products (e.g. Figure 3.11) are very expensive (US DOE, 2013a). The basic principle of an OLED is to pass electricity through one or more thin organic semi-conductive layers sandwiched between a layer of aluminium and a layer of transparent indium tin oxide. The whole sandwich is then attached to a sheet of glass or other transparent material known as a substrate. The current applied to the aluminium layer is conducted through the organic film which then emits light in different colours for different materials. Jacob (2009) observes:

> 'Developments based around flexible substrates would open a whole new concept of lighting. Imagine closing your curtains in the evening and they also become the primary or secondary light source to the room. Or maybe the OLED takes the place of the ... shades that would normally cover the lamps... The potential for OLEDs appear to be limited only by the designer's imagination, and as such are considered to be the next stage of evolution for ... lighting.

Internet connected lighting

Before novel lighting technologies such as OLEDs become options for general use, novel control systems are being developed that make use of the colour changing and other functions of LED lamps.

One such system launched by Philips in 2012 consists of sets of LED lamps controlled over the Internet using a smartphone. A *Which?* researcher conducted an initial test of the Philips 'Hue' starter pack and reported:

> For the first hour or so I had great fun adjusting the settings through the full spectrum of colours ... But the novelty soon wore off ... you end up realising that many of the modes make your living room look like a cheap night club... You can set different lighting 'moods' via your smartphone... You can also can dim or brighten them... Overall ... I enjoyed showing off all the different settings to my friends and finding the perfect lighting moods. [It] ... could be great for a party or to play with over Christmas. But on the downside ... it's expensive and you may need to buy extra bulbs to light up your room.
>
> (Which? Ltd., *2014d*)

Philips is adding new functions to its system and it is likely that other 'smart' home lighting systems will follow, for example, domestic versions of lights that switch on and off as people enter and leave rooms.

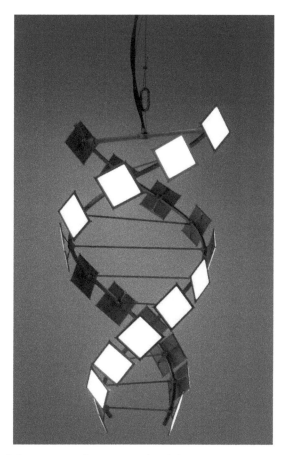

FIGURE 3.11 Philips DNA Helix concept chandelier using organic light emitting diode (OLED) panels, 2012

Sustainable lighting services

It is now becoming possible to obtain lighting as a service, thus overcoming some of the high upfront cost of purchasing new lighting technologies. One example is the lighting system that the National Union of Students (NUS) leases from Philips for its London office. Discussions between NUS and Philips led to the creation of a 'Pay per Lux' solution in which Philips retains responsibility for the performance of the lighting over a 15-year period and NUS pays for the energy used through a quarterly fee. Any replacements during the life of the contract will make use of the latest LED lamp technologies (Philips Lighting, 2014). This option of 'paying for light' rather than for lamps and fittings is being adopted first in commercial and business contexts, but it is possible that lighting could be one of the pay-per-use services (such as laundry, discussed in Chapter 2) that households will lease from service providers in the future.

The pattern of innovation

The evolution of electric lighting, unusually for a consumer product, has followed a generally divergent technological and design path. After the nineteenth century divergent experimentation with different designs of arc lamps and incandescent lamps – typical of the start of a new technology – the electric lamp converged on a single dominant design, the tungsten filament incandescent lamp. It then diverged to include fluorescent and halogen incandescent lamps in various shapes and sizes. It then diverged further to include CFLs and LED lamps, also in various shapes and sizes, before converging slightly with the withdrawal of GLS incandescent lamps from many countries. Other technologies and designs have been developed – for example, metal halide lamps and sodium lamps (widely used for street lighting) – but not normally used for domestic lighting and there are new lighting technologies such as OLEDs at the research stage or being developed to add to the divergent pattern. (A useful chart showing the evolution of electric lighting can be found at Edison Tech Centre (2013)).

This pattern of evolution has involved many people and organisations, large investments and efforts in scientific research, engineering and technological innovation. It has involved many inventors, patents and other innovations on which the creation and commercialisation of the lamps depended, including the introduction of production technologies to mass-manufacture affordable lamps. Despite the power struggles and patent disputes that lay behind the development of electric lighting, which sometimes influenced and delayed its progress, the drivers behind the many experiments and innovations usually had similar technical and commercial goals – increased operational efficiency and product life at acceptable cost, stimulated by the need to reduce environmental impacts and gain commercial advantage through innovation.

References

Advameg (2014) How products are made. Light bulb [Online]. Available at http://www.madehow.com/Volume-1/Light-Bulb.html (accessed April 2014).

Bijker, W.E. (1995) *Of Bicycles, Bakelites and Bulbs. Toward a theory of socio-technical change*, Cambridge, MA, MIT Press.

Consumers' Association (1975) 'Light bulbs', *Which?* April pp. 118–122.

Consumers' Association (1978) 'Which type of light bulb?' *Which?* September, p. 537.

Consumers' Association (1979) 'Getting lit up on the cheap', *Which?* October, p. 594.

Consumers' Association (1980) 'Fluorescent lighting', *Which?* February, pp. 96–99.

Consumers' Association (1981) 'Philips fluorescent light bulbs', *Which?* February, p. 110.

Consumers' Association (1985) 'Lighting', *Which?* March, pp. 120–124.

Consumers' Association (1993) 'Energy saving light bulbs', *Which?* May, pp. 8–10.

DECC (2013a) Energy Consumption in the UK, Domestic energy consumption in the UK between 1970 and 2012 [Online], London, Department of Energy and Climate Change. Available at https://www.gov.uk/government/uploads/system/uploads/attachment_data/file/65954/chapter_3_domestic_factsheet.pdf (accessed May 2014).

DECC (2013b) Energy Consumption in the UK. Domestic data tables update 2013, [Online] London, Department of Energy and Climate Change. Available at https://www.gov.uk/government/publications/energy-consumption-in-the-uk (accessed May 2014).

DeLair, R. (2013) Fluorescent Lamp Development: A comprehensive history covering the 1930s and 1940s [Online]. Available at http://www.edisontechcenter.org/fourescentlampdev.html (accessed May 2014).

EC (2009) Energy saving light bulbs [Online], European Commission. Available at http://ec.europa.eu/energy/lumen/professional/index_en.htm (accessed May 2014).

Edison Tech Center (2013) The electric lamp [Online]. Available at http://www.edison-techcenter.org/Lighting.html (accessed May 2014).

Encyclopedia Britannica (2013) Lamp: the Incandescent Lamp, [Online] The Editors of Encyclopedia Britannica. Available at http://www.britannica.com/EBchecked/topic/328821/lamp (accessed March 2014).

Jacob, B. (2009) 'Lamps for improving the energy efficiency of domestic lighting', *Lighting Research and Technology*, vol. 41, pp. 219–228.

Muis, H., Posthumus, A., Slob, A.F.L., van der Sluis, S.M. (1990) 'Environmental impacts of lighting: a product oriented approach.' Final Report 9008, Rotterdam, The Netherlands, Ministry of Housing, Physical Planning and the Environment.

Museum of Electric Lamp Technology (2013) [Online]. Available at http://www.lamptech.co.uk/index.html (accessed March 2014).

Open University (2014) 'Exploring innovation', T317 *Innovation; designing for change*, Block 1, Milton Keynes, The Open University.

Philips Lighting (2014) Philips delivers 'cradle to grave' lighting for NUS [Online]. Available at http://www.lighting.philips.co.uk/pwc_li/gb_en/projects/Assets/CaseStudy_NUS_UK.pdf (accessed May 2014).

PremiumLight (2014) PremiumLight Project [Online]. Available at http://www.premiumlight.eu/index.php?page=project-5 (accessed May 2014).

Smithsonian Institution (2008) Lighting a Revolution [Online]. Available at http://americanhistory.si.edu/lighting/index.htm (accessed May 2014).

Stokes, M., Crosbie T. and Guy, S. (2006) 'Shedding light on domestic energy use: a cross-discipline study of lighting homes', in Sivyer, E. (ed.) COBRA 2006: *Proceedings of the Annual Research Conference of the Royal Institution of Chartered Surveyors*, London, RICS.

US DOE (2013a) Life-Cycle Assessment of Energy and Environmental Impacts of LED Lighting Products, [Online]. US Department of Energy, Building Technologies Office, April http://apps1.eere.energy.gov/buildings/publications/pdfs/ssl/lca_factsheet_apr2013.pdf (accessed May 2014).

US DOE (2013b) Solid-State Lighting Research and Development: Multi-Year Program Plan, Building Technologies Office/Office of Energy Efficiency and Renewable Energy, U.S. Department of Energy DOE/EERE-0961, April [Online]. Available at http://apps1.eere.energy.gov/buildings/publications/pdfs/ssl/ssl_mypp2013_web.pdf (accessed April 2014).

US Energy Information Administration (2013) How much energy is used for lighting in the United States? [Online]. Available at http://www.eia.gov/tools/faqs/faq.cfm?id=99&t=3 (accessed May 2014).

Which? Ltd. (1999) 'The light fantastic', *Which?* October, pp. 30–32.

Which? Ltd. (2009) 'Low energy light bulbs rated', *Which?* November, pp. 28–29.

Which? Ltd. (2010) 'Test lab, Light bulbs', *Which?* July, pp. 30–31.

Which? Ltd. (2013a) 'Best Buys that last for years', *Which?* May, pp. 45–47.

Which? Ltd. (2013b) 'The brightest energy savers', *Which?* September, pp. 56–58.
Which? Ltd. (2014a) 'LED light bulbs are falling short', *Which?* February, p. 4.
Which? Ltd. (2014b) 'LED light bulbs outshine the rest', *Which?* September, pp. 67–69.
Which? Ltd. (2014c) 'LED bulbs in the spotlight', *Which?* July, pp. 69–71.
Which? Ltd. (2014d) Light bulbs: Philips Hue LED lighting system first look, [Online]. Available at http://www.which.co.uk/energy/energy-saving-products/guides/philips-hue-led-lighting-system-first-look/ (accessed May 2014).
Wilhite, H. (2001) ECEEE Position Paper on CFL in Europe, European Council for an Energy Efffficient Europe, March.

4
TELEVISION

Television is the consumer product and service that most people would be most reluctant to give up, only rivalled by mobile phones and the Internet (Pew Research, 2014). Electro-mechanical television was invented in the 1920s and all-electronic systems were developed in the 1930s, but the first television services were only offered to the public in London and New York just before the Second World War. Since then television has undergone a remarkable transformation.

At first, television involved watching a small cathode ray tube screen showing very few channels transmitted in black and white for a few hours per day. By the twenty-first century television had evolved into a system comprising a receiver with an increasingly large flat panel screen that provides 24-hour access to hundreds of TV channels in high-definition colour. Together with other equipment, there are options for cinema quality sound, recording programmes, playing games or pre-recorded disks, accessing catch-up TV services and the Internet or streaming online content to and from other devices including smartphones and tablet computers.

Television has penetrated to most parts of the world, becoming the dominant medium of mass communication and one of the most significant influences on society and culture.

Television technology

The earliest ideas for television ('seeing at a distance') originated in the late nineteenth century. In 1880 a French engineer, LeBlanc, proposed a scanning mechanism that would take advantage of the eye's temporary retention of a visual image. He envisaged a photoelectric cell that would view only one portion at a time of a picture to be transmitted. Starting at the upper left corner of the picture, the cell would scan to the right-hand side and then jump back to the left-hand side, only one line lower, until the entire picture was scanned. A receiver synchronised with the transmitter could then reconstruct the original image line by line. The concept of scanning – which, by limiting the amount of visual information to be transmitted at once, enabled a picture to be sent via a telephone line or by radio transmission – became the basis of television (Encyclopaedia Britannica, 2014). A key invention for early electro-mechanical TV was the Nipkow scanner of 1884, comprising a rotating disc with a spiral of holes. If a lens projects an image of a scene onto the disc, each hole in the spiral takes a 'slice' through the image which can be picked up as a pattern of light and dark by a photocell. If the photocell is made to control a light behind another Nipkow disc rotating at the same speed, the image will be reproduced line by line.

It wasn't until the early 1920s that the various components of television systems, including mechanical scanners, telephone and radio communications, the cathode ray tube and electronic valves were in place. Like other technological innovations, such as electric lamps (Chapter 3) and mobile phones (Chapter 6), television required the scientific basis and the invention and development of a number of prerequisite technologies before it could be implemented (Burns, 2004). The evolution of television technology in its formative years, as documented for example in a comprehensive book by Burns (1998), is a complex story of many dedicated people and organisations, inventions and patent disputes, commercially driven R&D programmes, government and business rivalries and collaborations.

Electro-mechanical systems

In the 1920s and 1930s experiments with both electro-mechanical and electronic television systems took place in Britain, America, France and Germany. In 1926 the world's first television system was demonstrated by a Scotsman, John Logie Baird, in his London laboratory. It was based on a Nipkow disc scanner and photocell which converted the scanned image of an object or person built up from 30 lines repeated 12.5 times per second into electrical signals for radio transmission to a 'Televisor' receiver with another photocell and rotating disc scanner that converted the signals into a flickering image on a 3-inch screen (Figure 4.1). From 1930 to 1935 regular television broadcasts were transmitted by the BBC on Baird's low-definition system and viewed by a few enthusiasts on Baird Televisors (BBC, 2014). Electro-mechanical television systems, based on similar principles and transmitted by wire or radio, were demonstrated by Bell Telephone Laboratories and AT&T in America and by individuals and companies in other countries (Burns, 2004; Genova, 2013).

Baird continued to develop electro-mechanical television systems after a rival all-electronic system had been chosen by the BBC for its public television service. Baird's demonstrations included cinema screen-sized colour television received by radio from another electro-mechanical transmitter at Crystal Palace 16 km away, and by 1941 Baird had created high-definition and stereo colour television using a combination of electronic cathode ray tubes and mechanical equipment (Burns, 2004).

FIGURE 4.1 Baird Televisor electro-mechanical television receiver with a rotating disc scanner and 3-inch screen (in the cavity on the right), 1930

Electronic systems

By the early 1930s Farnsworth Television and the Radio Corporation of America (RCA) in the USA and Electric and Musical Industries (EMI) in Britain were carrying out R&D on all-electronic television systems. Related work was going on in Germany and France. Such a system required the invention, design and development of two special types of cathode ray tube (CRT) – a camera tube to electronically scan the images to be transmitted and a display tube to show them on a television receiver. Three different camera cathode ray tube scanners were developed: Farnsworth's Image Dissector, then Zworykin's Iconoscope at RCA, and EMI's Emitron. At RCA Vladimir Zworykin developed a display cathode ray tube called the Kinescope. EMI also developed television receivers with display cathode ray tubes. These early electronic TV sets were used to receive broadcasts generated by both the new electronic and surviving electro-mechanical systems.

In November 1936 the BBC started the world's first public television service from its London station at Alexandra Palace. The BBC first had to decide on the best type of 'high-definition' television system, specified as an image of at least 240 lines at 25 frames per second. Hence a competition was arranged to test Baird's improved 240-line electro-mechanical system against EMI's 405-line all-electronic system with the Marconi Company providing the transmitter. Although Baird used Farnsworth's Image Dissector electronic camera as well as mechanical scanning in some tests, it soon became apparent that Marconi-EMI's all-electronic system with its manoeuvrable electronic camera was superior to Baird's cumbersome electro-mechanical equipment. Hence Baird's system was rejected after a few trials. The BBC's all-electronic television service began in January 1937 using the 405-line Marconi-EMI system (Burns, 2004). Between 1937 and 1939, when the Second World War interrupted broadcasting, the BBC transmitted two hours per day of films, studio programmes and outside broadcasts, including the King's coronation in 1937. These programmes could only be received in London and surrounding areas, but by 1939 some 19,000 TV sets, including electronic designs made by Baird's company (Figure 4.2), had been bought by the few who could afford them at about £86 (£5,000 in today's money) (Burns, 1998).

All-electronic public television started in America in 1939 when a programme was transmitted from the New York World's Fair to an estimated 100 to 200 television receivers. This was soon followed by 25 hours per week of television plays, films, sports and other programmes with 5–14-inch screen cathode ray tube TV sets on sale from four manufacturers at $200 to $600 ($3,400 to $8,000 today). Broadcasting was reduced to four hours per week from 1942 to 1945 after the US entered the War (Burns, 1998).

Post-war television

In June 1946 the BBC resumed TV broadcasting from Alexandra Palace for three hours per day using the pre-war 405-line system. By 1952 four new transmitting stations had extended the service to 28 million people outside London. This also encouraged the mass production of television sets, bringing down costs and making television more affordable. Following the BBC's highly popular broadcast of the Queen's Coronation in June 1953, television spread to 3.2 million British homes by 1954 (BBC, 2014).

FIGURE 4.2 Baird T-23 15-inch cathode ray tube mirror lid TV, 1938

The US had standardised on a 525-line system and resumed a full broadcasting service after the War, and by 1955 there were 36 million TV sets receiving many commercial and non-commercial stations. By then TV services using four different line standards had spread, at least on an experimental basis, to most industrialised and some developing countries, using the all-electronic systems pioneered by EMI in Britain and RCA in America (Burns, 1998).

Early colour television

An obvious improvement to black and white (monochrome) television was colour. Baird had demonstrated the first electro-mechanical colour television system in 1928, followed in 1929 by a similar system shown by Bell Labs in the USA. As mentioned earlier, in 1941 Baird demonstrated a mechanical/cathode ray tube (CRT) colour TV system. Baird's only competitor in the early 1940s was the American Columbia Broadcasting System (CBS), whose colour system used electronic cameras and TV receivers with synchronised rotating discs with coloured filters in front of their CRT screens. By then it was becoming clear that cumbersome electronic/mechanical colour TVs would not be suited to domestic use and would be incompatible with the existing monochrome system (Burns, 2004). Nevertheless, TVs that had a rotating colour wheel or drum in front of the CRT were produced for CBS's experimental broadcasts until the early 1950s before being abandoned (Genova, 2013).

All-electronic colour television owes its practical development to RCA Laboratories in America from 1945 to 1953. RCA's all-electronic colour TV, which was compatible with existing monochrome television, was the system approved by the US government (Magoun, 2004). The example of colour television shows that

FIGURE 4.3a RCA CT-100 15-inch colour television, one of the first American receivers with a colour CRT, 1954

for a consumer innovation to succeed, products that are practical for home use and compatible with established systems are needed.

By 1954 several US manufacturers were advertising colour TV sets, one of the first being RCA's 15-inch screen model at $1,000 ($13,000 today) (Figure 4.3a). By the end of 1954 RCA had produced a lower priced 21-inch colour TV and the technology began to take off (Genova, 2013). However, as happened before with different monochrome TV line standards, different colour TV standards were approved by governments around the world. The US NTSC standard was adopted by Japan, Canada and many Central and South American countries. In the 1960s an NTSC variant (PAL) developed in Germany was adopted in Western Europe, except by France which persuaded Russia and Eastern Europe to adopt its SECAM system (Magoun, 2004).

The modern era of television had begun.

From analogue to digital television

To keep the account of the development of modern television within limits, the remainder of this chapter will focus on domestic equipment and draw mainly on the products and innovations reported in *Which?* from 1958 to 2014. Although many television innovations first appeared in the USA or Japan, *Which?* reports on future developments before they reach the British market and so provides a good record of TV's technical and design evolution. At first, *Which?* television reports were published only every one to two years, but by the early twenty-first century, reports were appearing four or more times per year to keep up with the rapid changes in equipment and services.

Analogue television

The first *Which?* report on TV sets was in May 1960 (Consumers' Association, 1960). It noted that early post-war TV sets had 9, 12 or 14-inch screens, but by 1960 most British sets had 17-inch screens (measured diagonally) and cost £60 to £72 (about £1,200 to £1,400 today). These black and white TVs had 405-line cathode ray tubes (CRTs), valves and other components connected by wired and/or printed circuits. The sets could receive the two channels then available – BBC and ITV (Independent Television, the commercial service first launched to serve London in 1955).

In 1962 *Which?* reported that 70% of TVs in Britain, now typically with a 19-inch screen, were rented, partly due to price but also because 625-line TVs were in prospect capable of receiving the new UHF (ultra-high-frequency transmission) channels planned by the government (Consumers' Association, 1962). In 1964 *Which?* reported on the dual standard sets required to receive BBC2 on UHF 625 lines as well as the VHF (very-high-frequency) 405-line BBC1 and ITV channels. 19-inch screen valve TVs were still the most popular, priced from £69 to £135 (about £1,200 to £2,400 today) (Consumers' Association, 1964).

405-line VHF BBC1 and ITV broadcasts survived until 1985, when they were finally discontinued.

Colour television

Colour TV had been launched in Britain in 1967, but *Which?* reported that colour sets cost around £300 (about £4,750 today) and so it was best to rent (Consumers' Association, 1968). Nevertheless, a quarter of the population were soon viewing BBC2 in colour, with a rapid rise expected by the end of 1969 when BBC1 and ITV would also be transmitting in colour (Consumers' Association, 1969). In a 1972 survey *Which?* found that consumers were very satisfied with their colour TVs, saying that the extra cost over black and white was worthwhile. But the mostly British-made TVs were not very reliable, which was another reason for renting (Consumers' Association, 1972). By 1973 prices for a 22-inch set had fallen to around £250 (about £2,600 today). More TVs used reliable solid state electronics (transistors and integrated circuits) and so buying became a better option, especially if choosing the more reliable Japanese sets (Consumers' Association, 1973). A year later *Which?* tested 26-inch colour TV sets, the largest size then available, which could now connect to one of the domestic video cassette recorders (VCRs) just arriving on the UK market (Consumers' Association, 1974). Other innovations being introduced included remote controls and in-line CRTs in which the three colour electron guns were combined into one and, instead of dots, the picture was made up from vertical lines, as in Sony's innovative Trinitron tube (Figure 4.3b). Video cameras for recording home movies onto a VCR ('camcorders' had not yet been developed) and simple TV games consoles were also introduced from the late 1970s (Consumers' Association, 1977a).

FIGURE 4.3b Sony 18-inch colour TV with Trinitron cathode ray tube, c. 1970

Video recorders

Which? first reported on a video cassette recorder, the Philips N1502 VCR, in 1977 (Consumers' Association, 1977b). By the first full report on video recorders in 1979, *Which?* noted there were four incompatible domestic systems – JVC's VHS and Sony's Betamax (Figures 4.4a and 4.4b); Philips' VCR and Grundig's SVR – with machines costing £500 to £730 (about £2,200 to £3,200 today). They all used videocassette magnetic tapes and rotating recording heads to produce the fast tape speeds needed to record and play video and had timers to allow users to record programmes for viewing later – the main reason that people bought or rented one (Consumers' Association, 1979). By 1981 the VCR and SVR formats were being phased out and an improved V2000 format was introduced by Philips and Grundig, but it only lasted until 1988. Although Betamax and V2000 were technically superior in theory, the video 'format war' was eventually won by the Japanese VHS system. This was because VHS machines had already captured most of the US market, as they initially offered longer recording times than early Betamax and VCR machines and were cheaper, with the result that VHS had been licensed to more manufacturers than Sony's Betamax and the European VCR, SVR and V2000 systems. By 1981, *Which?* tests indicated that VHS machines produced pictures and sound as good as Betamax and V2000, could record for equal or longer, had equally good or better timers, and were priced similarly at £500 to £600 (about £1,700 to £2,000 today) (Consumers' Association, 1981a). Rental machines, pre-recorded films and 'camcorders' were also more widely

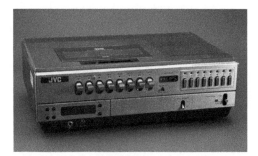

FIGURE 4.4a JVC Video Home System (VHS) video-recorder, 1976

FIGURE 4.4b Sony Betamax video-recorder, 1979

available in VHS than the two surviving formats, Betamax and V2000 (Consumers' Association, 1985). By 1987, *Which?* only tested VHS machines, Betamax was disappearing and V2000 had been discontinued (Consumers' Association, 1987a). This suggests that the longer recording and playing time and lower price of VHS machines, rather than the theoretically better technical specification of Betamax, gave VHS an edge with US consumers. Then the momentum behind VHS created by its many licensed manufacturers and users resulted in its worldwide success and the eventual demise of the other formats.

Other innovations

By 1981 the BBC and ITV teletext services (Ceefax and Oracle) and British Telecom's PRESTEL service, introduced in 1978–9, were fully operating. Both teletext services had 200 free pages of news, quizzes, recipes, etc. accessed via TV remote control buttons and PRESTEL provided 170,000 pages of paid-for information down the telephone line accessed via a keypad. Both systems could be slow to use, but they were primitive forerunners of the type of information now provided by the Internet (Consumers' Association, 1978; 1981b).

In the early 1980s *Which?* was anticipating multi-channel cable and satellite television services in Britain (Consumers' Association, 1981c) and reported on trials of 3D TV using 3D spectacles, which the magazine (correctly) anticipated would not prove especially popular with consumers (Consumers' Association, 1983a). Initially, satellite programmes were beamed at satellite dishes to redistribute to cable TV subscribers, but Sky and other channels broadcast directly to the home satellite dishes that had arrived by 1989 (Consumers' Association, 1987b, 1989). There was a boom in satellite TV subscribers in Britain from the mid-1990s, which meant that cable TV did not reach as many homes in Britain as it had in the US.

Personal TVs with 2-inch screens were made from the 1980s, but like 3D TV they were a premature technology that *Which?* advised consumers to avoid (Consumers' Association, 1984) and it had to wait until smartphones and tablet computers arrived before viewing TV and video on the move became practical.

Other innovations anticipated at the start of the 1990s included digital stereo sound, widescreen and high-definition TV and liquid crystal display (LCD) screens. Some manufacturers had already introduced widescreen 1250-line high-definition TV sets using the existing 625-line analogue PAL broadcasts; an example of making the best of existing technology before a 'proper' one becomes available (Consumers' Association, 1990).

Digital television

In the first decade of the twenty-first century, the transition towards today's TV technology took place. This involved the switch from analogue to digital television, high-definition television, integrated digital TVs with wide flat panel screens, personal video and DVD recorders.

Satellite broadcasting began the transition from analogue to digital television in the mid to late 1990s with the UK's analogue satellite broadcasts ceasing in 2001. Digital TV was a radical innovation comparable to the switch from electro-mechanical to all-electronic television. Its main advantage is that digital broadcasting allows more channels, with several programmes and other information on each channel, to be transmitted as compressed streams of data within the frequency bandwidths allocated to TV. This greatly increases the number of programme choices and enables services such as electronic programme guides and high-definition broadcasts to be provided (Miller, 2009). (Bandwidth is the width of the range, or band, of frequencies that an electronic signal uses when transmitted in a given medium.)

Terrestrial digital broadcasting, received via an aerial, came to the UK in 1998 with the launch of ONdigital, which eventually became Freeview. To view digital broadcasts consumers had initially to use a set-top box to convert the digital signal for viewing on an analogue TV, but the first integrated digital TVs (IDTVs) with built-in digital receivers soon followed. The transition to all-digital terrestrial television took place over different time periods in different countries using four different standards. For example, in Europe Digital Video Broadcasting-Terrestrial, DVB-T, replaced the analogue PAL standard. The first to complete the switchover was the Netherlands in 2006; the USA completed the switchover of its main TV stations in 2011 and the UK completed switching in 2012.

Digital Versatile Discs (DVDs) and DVD players, first marketed in Japan and the USA, were launched in the UK in 1998, but it would take some years before recordable DVDs and DVD recorders became affordable, so VCRs were still needed to record programmes. The LaserDisc, an earlier rival to videocassettes than DVDs, failed to take off as the player and discs were very expensive and could only hold one hour of film on each side (Which? Ltd., 1998). The LaserDisc (see Chapter 7, Figure 7.3) therefore failed in competition with VCRs for reasons similar to the earlier failure of Betamax in competition with VHS. With many more films broadcast via satellite and terrestrial digital television and on DVDs, home cinema kits with several loudspeakers, some providing surround sound, became more popular.

At first most TVs for viewing widescreen digital programmes still had cathode ray tubes (CRTs), even though flat panel liquid crystal display (LCD) and plasma TVs had entered the UK market in 2002–3. LCD TVs show an image when a light is shone through the screen's matrix of coloured liquid crystal cells, with TV signals controlling each cell (or pixel) letting varying amounts of colour through to build up a picture. In plasma TVs gas cells acting like tiny fluorescent lamps are sandwiched between two sheets of glass, with each cell emitting ultraviolet light that strikes red, green and blue spots on the screen which glow to make up the picture.

For some years *Which?* reported that most CRT TVs produced better pictures than LCD or plasma designs and were considerably cheaper (Which? Ltd., 2005) and so remained the choice of most consumers. By 2007 *Which?* found that flat screen LCD and plasma TVs could produce pictures as good as the best CRTs, although at a price. For example, a 32-inch Panasonic LCD cost £1,100 and a 40-inch Sony plasma cost £2,300 compared to £500 for a Best Buy 32-inch widescreen CRT model (Which? Ltd., 2007). By late 2008 *Which?* was only testing flat panel LCD and plasma TVs and prices had almost halved. Some of these sets could receive full high-definition television (HDTV), which was transmitted at first by digital satellite and cable TV (Which? Ltd., 2008). From 2009–10 digital terrestrial HDTV broadcasts started in Britain as analogue television broadcasts were switched off and more bandwidth became available.

High-definition television

High-definition television (HDTV) has a long history. Originally it meant the 405- or 525-line electronic television systems which replaced low-definition electro-mechanical television. The first 'proper' high-definition television system with a 1125-line analogue TV receiver was developed by the Japan Broadcasting Corporation in the 1980s with receivers produced commercially in the 1990s to receive a few HDTV broadcasts. But analogue HDTV never caught on worldwide because of incompatible standards promoted by different manufacturers and governments, its high bandwidth requirements and the high cost of HDTV sets, so its acceptance had to wait until the switch to digital television and the ability to compress the signal to save bandwidth (Sterling, 2004).

HDTV is one of the most significant innovations enabled by the digital switch, as it offers sharper and more lifelike pictures than standard definition TV. This is because the display is higher resolution, meaning there is more information to form the picture. Resolution is the number of pixels (picture elements) displayed on the screen. Full HD resolution is 1,920 x 1,080, meaning the screen has 1,920 pixels in each of 1,080 horizontal lines, scanned either interlaced (even and odd lines alternately called 1080i) or, for slightly higher definition, progressively (one after the other, called 1080p). This means that a HDTV screen has five times as many pixels (over 2 million) to display the picture as a standard resolution digital TV with 720 pixels in each of 576 lines. As well as HD broadcasts, an HDTV can be used to view other high-definition sources, including Blu-ray discs (a high-definition DVD format read using a blue rather than a red laser) and HD video streamed over the Internet. HDTV has also enabled larger screen sizes that still remain sharp.

Continuing innovation

A limitation of most TV sets has been their sound quality, and *Which?* found that TV sound had significantly deteriorated since the replacement of bulky CRT with slim flat panel TVs. For consumers who do not want a home cinema system with multiple speakers or to link to their hi-fi system, sound quality could be greatly improved by connecting the TV to a 'sound bar' placed under the set (Which? Ltd., 2012).

Light emitting diode (LED) TVs, which entered the UK market in 2009, are essentially the same as LCD sets, but the bulky back-light lamps are replaced by tiny LEDs. LED TVs are typically slimmer and more energy-efficient than LCD sets.

Two further innovations appeared around the same time as LED TVs; Internet-connected or 'smart' TVs and 3D TVs (capable of showing three-dimensional pictures). *Which?* was unimpressed by the early versions of smart TVs as they only offered limited Internet access via dedicated 'widgets' or apps, rather like early smartphones (see Chapter 6). 3D TV was also in its infancy, depending on the introduction of 3D broadcasts or 3D Blu-ray discs and viewers had to wear special glasses and buy an expensive large-screen TV to see a good 3D image (Which? Ltd., 2009). Although the 40–55-inch 3D TVs *Which?* tested worked well, most required viewers to wear heavy powered 3D glasses that blinked on and off. Given the limited amount of 3D content, consumers were advised to focus on 2D picture quality when buying (Which? Ltd., 2010a). Given its drawbacks it is not surprising that 3D was not the 'killer innovation' that attracted consumers to buy a new TV, although many models continue to offer 3D as an option.

Innovations that did attract consumers in large numbers were new ways of recording and catching up on missed TV programmes, given the increasing number to choose from. These technologies included hard disk personal video recorders (PVRs – Figure 4.5); DVD recorders with or without hard disks; and the start of online catch-up services such as the BBC's pioneering iPlayer. These digital technologies had by 2010 led to the demise of the analogue VCR and videocassettes (Which? Ltd., 2010b), as had happened before when CDs and digital downloading resulted in the end of audio-cassettes.

The new digital facilities for recording programmes; providing high-definition and smart TV; receiving terrestrial or satellite broadcasts and playing DVD and Blu-ray discs were usually provided via set-top boxes connected to the TV set. The consumer had to choose one or more boxes from the bewildering variety available to provide the digital services they wanted.

More of these digital facilities were built into the TVs introduced from 2010, making life simpler for the user. Many such integrated digital sets were improved

FIGURE 4.5 Sky+ personal digital video recorder (PVR). It can record two standard or high definition programmes on its 500GB to 2TB hard disk while showing a third programme, as well as access TV catch-up and Internet TV

smart TVs that could connect to the Internet via broadband to access TV catch-up services, watch YouTube videos, stream films, view Facebook etc. via pre-loaded 'apps' or, in some models, to browse the whole Internet. Depending on screen size and functions you could buy a 19-inch 'HD ready' TV (lower resolution than full HD) for £150 or a 55-inch full HD smart TV for nearly £3,000 (Which? Ltd., 2010a), showing that manufacturers were trying to cater for a full range of consumer market segments.

Many of the developments since 2010 in TVs and accessories like PVRs and Blu-ray players have been incremental innovations, such a smart TV with a camera for making Skype video-calls, or PVRs with hard disks capable of storing over 350 hours of high-definition TV.

A more significant development, however, has been the increasing number of online services, which has meant that viewing programmes, videos and films is no longer confined to a TV set, but may be streamed or downloaded to a mobile phone, tablet computer or PC (Which? Ltd., 2013). Some TV sets can also link wirelessly to smartphones or tablet computers to show the same or a different programme on a mobile screen or alternatively to 'cast' TV content from the smartphone or tablet for viewing on the TV's screen.

Television design

Pre-War television receivers

Before the Second World War television sets were either electro-mechanical or electronic devices. The Baird Company made both types; initially the electro-mechanical Televisor of 1929, designed to accommodate a fast-spinning scanning disc and displaying a small flickering picture at 30 lines per second, and connected to a separate radio set to receive the sound (see Figure 4.1). By 1936, when all-electronic television had been developed, the company made sets with 12–15-inch CRTs, mounted vertically because of their length, and an angled mirror to view the picture (see Figure 4.2). Such TV sets were expensive items of furniture, often with fashionably styled veneered wooden cabinets and facilities such as radios and record players, even cocktail cabinets, for the few who could afford them (Bennett-Levy, 1993). For example, the Marconiphone 701 had a floor-standing Art Deco walnut cabinet housing valve receivers for picture and sound signals and a 9-inch vertical CRT showing a 10 x 8-inch picture on a mirror lens. It cost £80 in 1936 (about £5,000 today). Lower priced TVs with smaller screens could have horizontally mounted CRTs, like the 5-inch screen Pye table top set of 1938, which required a separate radio for sound and cost £25 (about £1,500 today). In the US RCA-Victor made pre-war TV receivers with 'streamlined' walnut cabinets and horizontal CRT screens ranging in size from approximately 3 x 4 inches to 5 x 7 inches, priced in 1939 at $200 to $450 (about $3,300 to $7,500 today) (Figure 4.6).

FIGURE 4.6 RCA TRK-9, a TV receiver with a 9-inch diagonal cathode ray tube and radio housed in a 'streamlined' veneered wooden cabinet, 1939

Post-War cathode ray tube TV sets

In the immediate post-war period, TV sets were still either large items of wooden furniture (Figure 4.7a) or small-screen table-top models (Genova, 2013). To reduce costs cabinets were sometimes made from moulded plastics; an early example was the 1948 Bush TV12 (Figure 4.7b). By the mid-1950s the design of typical black and white TV sets had converged onto a dominant design, comprising a deep rectangular cabinet to accommodate the long cathode ray tube with the screen, speakers and/or controls filling most of the front and usually designed to sit on a table or stand (Figure 4.7c) (Bennett-Levy, 1993).

With the exception of a few unusual models such as the 1970 spherical Keracolor set, TV sets of the late 1960s onwards – black and white, colour or portable – were of this dominant design. In this period component innovations, such as solid state electronics and Sony's Trinitron colour tube (see Figure 4.3b) were incorporated. At the same time manufacturers introduced incremental improvements such as push button or touch controls, squarer screens, and designs for economic manufacture and easier servicing.

Until the introduction of flat panel screens in the early twenty-first century, the dominant design of TV sets hardly changed, as a deep cabinet was still needed for the CRT, even if the rest of the electronics was solid-state. The main difference by the 1990s, when remote controls had displaced the set's controls, was that the screen usually filled almost the whole front of the cabinet, which was typically made of grey or black plastic and had a more curved form than earlier (mainly wooden) sharply rectangular TV cabinets. Most large TVs came with stands with shelves to accommodate accessories such as a VCR or satellite TV receiver

FIGURE 4.7a Baird 12-inch 405-line black and white CRT TV, 1949. It was available as the Townsman, for reception close to the transmitter, and the Countryman for fringe areas

FIGURE 4.7b Bush TV12 with 9-inch 405-line black and white CRT and Bakelite case, 1948

FIGURE 4.7c GEC TV set with 14-inch 405-line black and white CRT, 1955

(Figure 4.8). Screens also increased in size to 28 inches or more and in 16:9 widescreen, rather than the old 4:3 format, as more programmes were broadcast in widescreen. Digital TV by satellite or terrestrial broadcast arrived in the UK by 1998, but apart from being able to link to a digital set-top box or having an integrated digital tuner, the external design of TVs hardly changed.

FIGURE 4.8 Sony 'big box' 32-inch widescreen cathode ray tube colour TV, c. 2005

Other television equipment

Most new design went into the equipment that went with the TV, such as VCRs and set-top satellite boxes. Many consumers found operating this new equipment difficult and so human factors research led to improved user interfaces, such as the VideoPlus system for programming VCRs by entering a code printed in TV guides into a special remote control (Consumers' Association, 1992a). After the arrival of digital TV, electronic programme guides enabled users to find programmes more easily, view or record them, or even scroll back over past programmes and view them via online catch-up services.

Flat panel TV receivers

Flat panel liquid crystal display (LCD), plasma and light emitting diode (LED) screens enabled TV set design to be radically changed from a bulky black or grey plastic cabinet to a shallow box framing a rectangular flat screen (Figures 4.9a, 4.9b). Such sets could be wall-mounted.

As the technologies evolved, screen sizes became steadily larger. In 1945 a 12-inch screen, in 1960 a 21-inch screen, and until about 2005 a 'big box' TV with a 32-inch CRT screen, were considered large. By 2014 a large flat panel TV had a 40–55-inch or bigger screen. And as each size of TV was widely adopted its price fell; for example, the UK price of a 55-inch LED TV fell from about £3,000 in 2010 to between £1,000 and £1,800 by 2014. Some consumers bought video projectors to project an even bigger picture when connected to a digital TV, set-top box or Blu-ray player. However, *Which?* warned consumers tempted to buy a video projector that they would need a very big room to sit far enough away to view it comfortably (Which? Ltd., 2014b).

FIGURE 4.9a Panasonic 32-inch screen flat panel LCD TV, 2005

FIGURE 4.9b Samsung 40-inch screen LED Smart TV, 2013. Note the narrow screen border compared to Figure 4.9a

Environmental impacts and regulation

Concern about the environmental impacts of television did not become a significant factor in TV equipment design and marketing until the early twenty-first century.

Television energy use

Most life cycle assessment (LCA) studies show that, like other energy-using products, the main environmental impact of television equipment arises from its energy consumption and greenhouse gases emissions when in use (WRAP, 2010). Although individual items of TV equipment are not major energy users, their total energy use is significant. In the EU, for example, TV equipment electricity use in 2007 was estimated as 60 billion kWh per year (Intertek, 2014).

UK data shows that the electricity used by TVs, set-top boxes and DVD/VCR equipment increased by 70% from 1990–2009, when it accounted for a quarter of household electricity use, more any other type of domestic electrical equipment (EST, 2011). It has been calculated that the electricity used by TV equipment generated about 300 kg carbon equivalent greenhouse gas emissions per UK household per year; a small but growing percentage of the average household's carbon footprint (Druckman and Jackson, 2010). TV equipment is one of the few classes of household product whose energy use has risen rapidly and is expected to continue rising, despite innovations such as more energy-efficient LED TVs. This is partly due to more items of equipment per household and partly to increasing screen sizes and new services such ultra-high-definition TV (EST, 2011).

Early electronic TV sets had valves and a CRT and required fair amounts of power; the Marconiphone 701 9-inch screen TV of 1936, for example, used 260 watts. The first mention of TV energy use in *Which?* was in 1976, when it reported that the most energy-efficient 18-inch screen sets used 68 watts, while the least efficient at 222 watts used over three times as much. *Which?* commented that if everyone had the more energy-efficient TVs, it would save a worthwhile amount of electricity (Consumers' Association, 1976). In 1981 *Which?* reported that 26-inch colour TVs, the largest size then available, used 73–191 watts and if on for 5 hours per day would use 133–333kWh per year (Consumers' Association, 1981b), while an average VCR used about 118 kWh per year if operating 10 hours per week and left on continuous standby (Consumers' Association, 1981a).

From the 1980s to the early 2000s, after which they began to be displaced by flat panel TVs, there was not much change in the energy-efficiency of CRT TVs. However, during that time there was growing concern about the amount of electricity used when equipment was left on standby. In 1992 *Which?* reported that an average CRT TV used half the electricity on standby as when switched on, although some models used much less (Consumers' Association, 1992b). An additional problem was the introduction of set-top digital receivers, some of which had to be kept on constant standby, using three to four times the electricity of an average TV on standby (Which? Ltd., 1999). Although a relatively minor problem in the context of total energy use, it stimulated regulatory actions, such as the 1999 International Energy Agency's '1-watt initiative', which led to the average new TV set's standby consumption falling from about 5 watts to 1 watt or less. However, by 2002 most set-top boxes still used 6–8 watts on standby, although one model used less than 0.5 watt (Which? Ltd., 2002).

In the 2000s flat panel LCD and plasma TVs began to displace CRT models. At first *Which?* found that LCD TVs used slightly more electricity than CRT models, but plasma TVs used two to three times as much electricity as CRT and LCD sets (Which? Ltd., 2004). By 2009, 40-inch LCDs were reported to use 99–223 watts and plasma models were using 208–434 watts; i.e. the plasma models required from the same to four times the electricity of the LCD TVs. All models used well under 0.5 watt standby power and some had sensors that automatically dimmed the backlight in low light conditions (Which? Ltd., 2009).

At the same time, more energy-efficient LED TVs began to displace LCD and plasma screens. *Which?* online compared typical 42-inch TVs of each type, reporting that the LED TV used 64 watts, the LCD set 107 watts and the plasma TV 195 watts (Laughlin, 2014a). Other equipment was also designed to be increasingly energy-efficient. For example, in the five years to 2014 the power requirement of efficient DVD recorders and PVRs had halved to under 15 watts in use and to under 0.5 watts on standby (Which? Ltd., 2014a).

The shift towards more energy-efficient TVs was partly driven by environmental legislation and standards that specified steadily increasing levels of energy-efficiency. The regulations include the US voluntary Energy Star scheme, which from 2008 labelled TVs of different sizes according to maximum power requirements (Energy Star, 2012). From 2009 the EU Directive for Energy-related Products (ErP) for televisions specified maximum power consumption levels for simple set-top boxes and per unit of screen area for TVs (Intertek, 2014). Mandatory EU energy labelling of televisions was introduced in 2011 with energy use per unit of screen area rated on an A to G scale (Figures 4.10a and 4.10b). The scale was revised from 2013 to include A+ (a 55-in HDTV with this rating uses about 76 watts) and in future to include A++ and A+++ classes.

There are now 47–55-inch colour LED HDTVs available that use the same power as the most efficient 18-inch monochrome CRT sets of 40 years ago (Energy Star, 2012). All the major TV equipment manufacturers have ambitious

FIGURE 4.10a Samsung small 22-inch (54 cm) screen LED Smart High Definition TV, 2014

FIGURE 4.10b The Samsung 22-inch TV's 'A' energy-efficiency rating, power and annual energy use on its EU Energy Label

sustainability policies. Sony, for example, calculates the life-cycle carbon emissions of its TVs and disc players and has introduced a product development process aimed at creating more energy- and resource-efficient ranges (Sony, 2013). Nevertheless, it can be argued that they are failing to reduce their products' energy use as much as possible by developing ever higher performing products. More generally, this suggests that the effectiveness of regulatory attempts to reduce the environmental impacts of consumer goods is limited while manufacturers design ever larger, more powerful and feature-laden products, which consumers are willing, or can be persuaded, to buy.

Materials and waste

Apart from the emissions from its energy use, TV equipment has other environmental impacts. Some LCA studies suggest that, depending on the energy supply mix, the global warming and air pollution impacts of manufacturing a TV from materials, including glass, steel, aluminium, copper and plastics, are greater than those of the energy during its use (Huulgaard, Dalgaard and Merciai, 2013). In response, the major TV manufacturers' sustainability policies include strategies for reducing the environmental impacts of materials and manufacture.

However, it was concern about the increasing volumes of electrical and electronic equipment discarded (2 million TVs thrown away each year in the UK alone) that led to legislation, notably the EU's WEEE (Waste Electrical and Electronic Equipment) and RoHS (Restriction of Hazardous Substances) Directives introduced in 2003. This legislation restricts the use of hazardous substances (e.g. lead, cadmium and mercury) in electronic equipment and sets targets for the collection and recycling of such equipment. However, despite the legislation and an electronics recycling industry, much TV equipment is still going to landfill. In addition, the amounts discarded are increasing, not least due to the rapid innovation in TV equipment as manufacturers compete to introduce new technologies and features to persuade consumers to replace products before the end of their life. Samsung, however, is making some attempt to address this issue by extending the life of its smart TVs with plug-in modules to enable consumers to upgrade the TV's processor and software (Samsung, 2014).

Social influences and impacts

Raymond Williams' pioneering study 'Television: Technology and Cultural Form' (1974, 1990) criticised the popular and simplistic view that television is an innovation arising from scientific and technological research that 'has altered our world'. He argued that television did not arise as a result of inevitable 'technological determinism' with unforeseen social consequences, but emerged from the interaction of science and technology with socio-economic and political forces. Others have documented the government controls and regulations, international and commercial rivalries and alliances, business interests, creative teams and many other non-technical factors that shaped the evolution of television technology (Burns, 1998). Like lighting (Chapter 3) it can be argued that television technology has been 'socially shaped'.

From work such as that of Williams, the academic subject of television studies developed (Geraghty and Lusted, 1998). It is not possible to summarise the work of this whole discipline, or the many discussions and debates concerning the social, political and cultural influences on and impacts of television, for instance whether television has increased violence, or passivity, in society, or has beneficial or harmful educational effects. In any case, television studies and the debates concerning television focus mainly on its content, financing and organisation rather than its design and technology.

Nevertheless, innovation has affected the organisation and content of programmes, their cultural impact and how people view TV. Television started before the Second World War as something for a few wealthy innovators and enthusiasts. In Britain in the early 1950s only about 15% of households had a TV, on which they could watch one black and white channel broadcasting five hours per day. By the early 1980s mass production of sets meant that 95% of British households could afford to rent or buy a set and there were four channels broadcasting about 50 hours per day of colour programmes (Consumers' Association, 1983b). Among the studies of the impacts of TV technology on consumers is a five-year survey of the viewing habits of 450 representative members of the British population. It found that widespread adoption of VCRs as an innovation in the 1990s had brought 'almost revolutionary changes to people's ability to schedule their own preferred material'. But the survey also found that most television was still watched at broadcast times with programmes such as the news and soap operas marking fixed times in people's schedules when family members could both bond and argue (Gauntlett and Hill, 1999).

Innovation in television technology is continuing to have social impacts. For example, the public service model of television exemplified by the BBC, and the shared culture of people viewing the same few programmes, has been challenged by the availability of hundreds of free and subscription channels enabled by satellite and digital TV. The innovations of personal video recorders, TV catch-up services, and online video streaming and TV via smartphones and tablet computers have had significant impacts on viewing habits, with fewer people now sitting down to watch a scheduled programme. With several ways of viewing and numerous channels, many are watching more television than before – in Britain, for example, over four hours per day in 2013, up from 3.5 hours in 2006 (Which? Ltd., 2013). However, other screen-based activities, such as accessing social media and gaming on smartphones and tablet computers, are eating into TV viewing time, especially among the young (see Chapter 6). But smartphones, tablets and lightweight video cameras have also enabled a new type of 'television' in which almost anyone can produce a video and upload it to YouTube for the world to view online.

Future developments

By late 2014, new display technologies were being introduced by TV manufacturers to provide even higher definition pictures and larger screens of up to 84 inches or more. These innovations included '4K' ultra-high-definition TVs, OLED (organic light-emitting diode) displays and curved screens (Figure 4.11). At the same time plasma TVs were being phased out as offering no performance advantages over existing LED TVs as well as being less energy-efficient (Which? Ltd., 2014b). Like TVs with cathode ray tubes previously, plasma TV is an example of a technology that survived for a period before being displaced by an alternative, and eventually superior, technology.

Ultra-high-definition television

A 4K LED TV displays a picture at 3,840 horizontal pixels by 2,160 vertical pixels or lines (totalling over 8 million pixels), four times as many as full HD (1,920 x 1,080). As there are so far no 4K broadcasts and very few 4K Blu-ray discs, the TVs electronically 'up-scale' standard or high-definition pictures to ultra-high-definition. Initial *Which?* tests of 4K sets reported impressive up-scaled picture quality, but did not consider the improvement over full HD worth their extra cost (Which? Ltd., 2014b). Nevertheless, *Which?* believed that 4K was the future of TV despite many technical barriers, including the high bandwidth required to broadcast 4K and very fast Internet connections to stream or download 4K online content. However, new file compression technologies and the agreement in 2014 of a European standard (DVB_UHDTV) could make 4K widely available in the future (Which? Ltd., 2014c).

FIGURE 4.11 Samsung 55-inch '4K' Ultra-High Definition curved screen LED smart TV

OLED screens

Organic light-emitting diode screens are a new display technology being introduced for both HD and 4K TVs (OLED is also a new technology for lighting and mobile phones, see Chapters 3 and 6). With OLED technology, the screen's pixels light themselves rather than needing a backlight, offering the deeper blacks of plasma screens and more vibrant colours. It also means that OLED TVs can be ultra-thin and flexible, allowing easier production of TVs with curved screens, which manufacturers claim provide a more immersive viewing experience (Laughlin, 2014b). Many reviewers including *Which?*, however, were unconvinced of the advantages of curved screen sets (Which? Ltd., 2014d). However, in the near future OLED offers the prospect of TV screens that roll up.

As with every previous innovation from colour to digital and HDTV, OLED screen TVs are expensive, at over £2,000, when first launched then coming down in price to become more generally affordable.

The pattern of innovation

Television is a striking example of continuous and accelerating innovation, with one generation of technology displacing an earlier one and many new products and services, such as colour TV, digital broadcasting, personal video recorders, high-definition and online TV being introduced (Genova, 2013; Santos, 2013). This pattern of innovation has been driven as much by social, political, commercial and economic forces as by science and technology. Arguably, television is an example of a stream of socially shaped technological innovations over which the consumer has had relatively little influence, apart from wanting improved reliability and lower prices and deciding whether or not to buy. The adoption of TV equipment has, however, followed a consistent pattern. Each new TV technology or device is very expensive when first introduced and so only bought by wealthy or enthusiastic consumers. Then prices fall and, in most cases, the technology is adopted more widely, usually displacing earlier technologies.

There has been less innovation in the design of the cabinets or cases that house the changing technology. In the initial phase of technological development, television receivers had various forms as the electro-mechanical and electronic systems competed for dominance. By the mid-1950s the familiar dominant design of a box-shaped CRT TV set had been established. Subsequently digital CRT televisions did not look much different from their analogue predecessors. And once the flat panel form had been established in the early twenty-first century, the evolution from LCD and plasma to LED displays had relatively small impacts on external design apart from the thickness of the set and the possibility of curved screens. However, OLED technology may enable more radical changes in TV design, such as bendable or rollable screens.

References

BBC (2014) History of the BBC [Online]. Available at http://www.bbc.co.uk/history-ofthebbc/resources/tvhistory/index.shtml (accessed May 2014).

Bennett-Levy, M. (1993) *Historic Televisions and Video Recorders*, Musselburgh, MBL Publications.

Burns, R.W. (1998) *Television. An International History of the Formative Years*, London, The Institution of Electrical Engineers.

Burns, R.W. (2004) 'Television: Beginning ideas; Color, Electro-mechanical, Electro-mechanical systems', in Hempstead, C. (ed.) *Encyclopaedia of 20th Century Technology*, Vol. 2, London, Routledge, pp. 818–819; 823–824, 830–832.

Consumers' Association (1960) 'TV sets', *Which?* May, pp. 96–103.

Consumers' Association (1962)' TV rent or buy?' *Which?* November, pp. 323–335.

Consumers' Association (1964) 'TV sets', *Which?* July, pp. 208–215.

Consumers' Association (1968) 'Colour TV', *Which?* January, pp. 13–17.

Consumers' Association (1969) 'More about colour TV', *Which?* May, pp. 148–150.

Consumers' Association (1972) 'Colour TV rentals, reliability and servicing', *Which?* September, pp. 279–286.

Consumers' Association (1973) 'Colour TV sets 18 in–22 in', *Which?* September, pp. 265–270.

Consumers' Association (1974) '26 inch colour TVs', *Which?* October, pp. 297–301.

Consumers' Association (1976) '18-in colour TV', *Which?* January, pp. 7–9.

Consumers' Association (1977a) 'Inside story: TV games', September, p. 526.

Consumers' Association (1977b) 'Inside story: Video cassette recorders', October, p. 531.

Consumers' Association (1978) 'Inside story: teletext a new service', *Which?* January, p. 3.

Consumers' Association (1979) 'Video cassette recorders', *Which?* July, pp. 417–423.

Consumers' Association (1981a) 'Video recorders', *Which?* March, pp. 196–201.

Consumers' Association (1981b) '26-in colour TVs', *Which?* January, pp. 3–8.

Consumers' Association (1981c) 'Inside story: Tomorrow's TV', *Which?* May, p. 314.

Consumers' Association (1983a) 'Inside story: 3D TV', *Which?* September, p. 388.

Consumers' Association (1983b) 'The viewers' view', *Which?* September, pp. 494–497.

Consumers' Association (1984) 'How to choose your new TV', *Which?* January, pp. 27–31.

Consumers' Association (1985) 'Video recorders', *Which?* February, pp. 72–75.

Consumers' Association (1987a) 'Video recorders', *Which?* June, pp. 286–289.

Consumers' Association (1987b) 'What's on TV?' *Which?* January, pp. 12–13.

Consumers' Association (1989) 'Satellite television', *Which?* February, pp. 79–81.

Consumers' Association (1990) 'TV into the 1990s', *Which?* February, pp. 111–113.

Consumers' Association (1992a) 'Video recorders', *Which?* December, pp. 42–45.

Consumers' Association (1992b) 'Television on test', *Which?* January, pp. 19–21.

Druckman, A. and Jackson, T. (2010) 'An exploration into the carbon footprint of UK households', RESOLVE Working Paper 02-10, Guildford, Centre for Environmental Strategy, University of Surrey.

Encyclopaedia Britannica (2014) Television (TV), [Encyclopaedia Britannica Online Academic Edition]. Encyclopædia Britannica Inc. Available at http://www.britannica.com.libezproxy.open.ac.uk/EBchecked/topic/1262241/television-technology (accessed May 2014).

Energy Star (2012) Product retrospective. Televisions, US Environmental Protection Agency [Online]. Available at http://www.energystar.gov/ia/products/downloads/TVs_Highlights.pdf (accessed July 2014).

EST (2011) The elephant in the living room: how our appliances and gadgets are trampling the green dream [Online]. London, The Energy Saving Trust, October. Available at http://www.energysavingtrust.org.uk/Publications2/Corporate/Research-and-insights/The-elephant-in-the-living-room (accessed July 2014).
Gauntlett, D. and Hill, A. (1999) *TV living. Television, Culture and Everyday Life*, London, Routledge.
Genova, T. (2013) Television History – The First 75 Years [Online]. Available at http://www.tvhistory.tv/index.html (accessed June 2014).
Geraghty, C. and Lusted, D. (eds) (1998) *The Television Studies Book*, London, Arnold.
Huulgaard, R.K., Dalgaard, R. and Merciai, S. (2013) 'Ecodesign requirements for televisions—is energy consumption in the use phase the only relevant requirement?' *International Journal of Life Cycle Assessment*, vol. 18, pp. 1098–1105.
Intertek (2014) Analysis of the Televisions Implementing Measure, Intertek, Boxborough, MA [Online]. Available at http://www.intertek.com/electrical/erp-directive/televisions/ (accessed February 2015).
Laughlin, A. (2014a) Advice Guide. LED vs LCD vs plasma TV [Online]. Available at http://www.which.co.uk/reviews/televisions/article/buying/led-vs-lcd-vs-plasma-tv- (accessed February 2015).
Laughlin, A. (2014b) Advice guide. What is OLED TV? [Online]. Available at http://www.which.co.uk/reviews/televisions/article/buying/what-is-oled-tv (accessed February 2015).
Magoun, A.B. (2004) 'Television: Color, Electronic', in Hempstead, C. (ed.) *Encyclopaedia of 20th Century Technology*, Vol. 2, London, Routledge, pp. 825–827.
Miller, M. (2009) Analog versus Digital TV: What's the Difference? [Online]. Available at http://www.quepublishing.com/articles/article.aspx?p=1245329&seqNum=4 (accessed July 2014).
Pew Research Center (2014) Pew Research internet project, February [Online]. Available at http://www.pewinternet.org/2014/02/27/summary-of-findings-3/ (accessed February 2015).
Samsung (2014) Eco-design [Online]. Available at http://www.samsung.com/us/about-samsung/sustainability/environment/eco_products/eco_design.html (accessed July 2014).
Santos, A. (2013) The evolution of live TV, TechGeeze, 8 June [Online]. Available at http://www.techgeeze.com/2013/06/evolution-of-live-tv.html (accessed July 2014).
Sony (2013) Reducing the Environmental Impact of Products and Services [Online]. Available at http://www.sony.net/SonyInfo/csr_report/environment/products/index.html (accessed July 2014).
Sterling, C.H. (2004) 'Television, digital and high definition systems', in Hempstead, C. (ed.) *Encyclopaedia of 20th Century Technology*, Vol. 2, London, Routledge, pp. 827–829.
Which? Ltd. (1998) 'What goes around', *Which?* June, pp. 32–34.
Which? Ltd. (1999) 'Ready set go', *Which?* April, pp. 26–29.
Which? Ltd. (2002) 'Digital Daze', *Which?* October, pp. 30–32.
Which? Ltd. (2004) 'Battle of the big screens', *Which?* May, 2004, pp. 30–34.
Which? Ltd. (2005) 'Widescreen TVs', *Which?* May, pp. 50–52.
Which? Ltd. (2007) 'Flat-panel progress', *Which?* April, pp. 38–41.
Which? Ltd. (2008) 'Screen savers', *Which?* September, p. 55.
Which? Ltd. (2009) 'Time for hi-tech television', *Which?* August, pp. 30–35.
Which? Ltd. (2010a) 'Test lab: 3D televisions', *Which?* September, pp. 48–51.
Which? Ltd. (2010b) 'Test lab: DVD recorders', *Which?* April, pp. 47–49.
Which? Ltd. (2012) 'The best TV sound for all budgets', *Which?* December, pp. 46–49.

Which? Ltd. (2013) 'How to watch TV (without a TV)', *Which?* June, pp. 33–37.
Which? Ltd. (2014a) DVD and PVR the pros and cons [Online]. Available at http://www.which.co.uk/technology/tv-and-dvd/guides/how-to-buy-the-best-dvd-recorder/dvd-or-pvr--the-pros-and-cons/ (accessed July 2014).
Which? Ltd. (2014b) 'Score yourself a TV bargain', *Which?* April, pp. 50–54.
Which? Ltd. (2014c) 'Next generation TVs', *Which?* January, pp. 48–49.
Which? Ltd. (2014d) Review. Samsung UE48H8000 [Online]. Available at http://www.which.co.uk/reviews/televisions/samsung--ue48h8000/review (accessed July 2014).
Williams, R. (1974) *Television: Technology and Cultural Form*, London, Fontana (Second edition London, Routledge, 1990).
WRAP (2010) *Environmental assessment of consumer electronic products*, Waste Resources and Action Programme, UK, May.

5
VACUUM CLEANERS

Today there are very few homes in industrialised countries without a vacuum cleaner. But like other domestic appliances that aimed to save labour, such as washing machines (Chapter 2), the spread of vacuum cleaners into general use was surprisingly slow. Vacuum cleaners appeared on the British and US markets at the start of World War One, but it was not until 1942 that half of US households, and not until 1955 that half of English ones, owned one (Bowden and Offer, 1994). Part of the reason was that in 1930s Britain a vacuum cleaner, even bought on hire purchase, was beyond the means of most people. In any case most homes, even if they had an electricity supply to power lights and a radio, did not have the necessary power sockets for a vacuum cleaner. Even in the much wealthier US, consumers preferred to buy a radio to entertain the whole family rather than a vacuum cleaner to ease the task of the housewife or servant who cleaned (Bowden and Offer, 1994). Between the World Wars vacuum cleaners were mainly bought for well-off middle-class households, at least partly because of the 'servant problem' to help move from having live-in servants to employing a household help. For the middle-class woman owning a vacuum cleaner, even if the 'daily' used it, it was also seen as symbolising modernity (Ryan, 2006).

After the Second World War vacuum cleaner sales gradually increased until by the early 1970s almost every household had one. Today the Northern European, US, Australian and Japanese markets are saturated and most consumers only buy a new cleaner when their existing one breaks down. Even in countries such as Spain, Portugal, Italy and Turkey where hard floors are more common than fitted carpets, vacuum cleaners are found in 80–90% of households, while in newly industrialised countries such as India, China and Brazil about a third of homes now have one (Floating Path, 2013).

Vacuum cleaner technology and design

Before vacuum cleaners, rugs and carpets were taken outside, hung over a wall or line and hit with a carpet beater to drive out as much dirt as possible. This arduous task was only done occasionally. Furnishings and hard floors had also to be dusted or swept by hand. The vacuum cleaner – a device that uses an air pump to create a partial vacuum to suck up dust and dirt – evolved, like the washing machine, from various manually-operated cleaning devices (Figure 5.1).

Early electric vacuum cleaners

In common with the other consumer products discussed in this book, the evolution of the vacuum cleaner involved very many individuals and companies, attempts to apply a number of different technical principles and the development of numerous designs of varying performance and commercial success. A detailed account of this highly complex history is provided in a book by Gantz (2012), former head of industrial design at Hoover and at Black and Decker. In this chapter I will only focus on what are generally considered to be the most significant innovations and designs (see e.g. Bellis, 2015).

In 1869, a Chicago inventor, Ives McGaffey, patented one of the first devices that used a vacuum system; a 'sweeping machine' he named the Whirlwind. The machine used a hand-cranked fan to generate suction and so was hard to operate because the user had to turn the crank while pushing the machine over the floor. Soon after, in 1876, Melville and Anna Bissell, owners of a Michigan china shop, developed the first commercially successful carpet sweeper. Their invention was spurred by the need to sweep pieces of china packing-crate straw from their carpets – a classic case of 'necessity' being 'the mother of invention'. When pushed along, the wheels of Bissell's carpet sweeper turned brushes mounted under a dust container, thus sweeping dirt into the container (Figure 5.1 middle left).

One of the first powered vacuum cleaners was invented by a British engineer, Hubert Booth, and patented in 1901. Booth got his idea after seeing a demonstration of an American-made powered cleaning machine which, rather ineffectively, blew dust into a collecting box. Booth argued that a suction machine should work better and, after mulling over ('incubating') the problem, demonstrated the principle by placing a handkerchief over the cover of a restaurant armchair and sucking hard, which left a patch of dirt on the handkerchief. Booth then developed a machine based on the suction principle, but which was nothing like a modern vacuum cleaner. It consisted of a horse-drawn carriage on wheels containing an oil engine-driven pump, which was parked outside the building to be cleaned. The rooms were vacuumed via long hoses fed through the windows (Figure 5.2). Booth's 'Puffing Billy' was given royal approval after it had successfully cleaned the carpets in Westminster Abbey for Edward VII's Coronation in 1902. Society women invited their friends to 'vacuum parties' to watch the dust being sucked from their carpets through transparent sections of hose connected to one of Booth's machines (van Dulken, 2002).

FIGURE 5.1 Bissell carpet sweeper late 1870s (middle left) and various designs of late nineteenth and early twentieth-century bellows or bicycle pump-operated manual vacuum cleaners

FIGURE 5.2 Booth's 'Puffing Billy' horse-drawn vacuum cleaner, 1901. The machine was parked outside the house to be vacuumed and the cleaning hoses fed through the windows. It was originally powered by an oil engine inside the carriage; an electric motor was substituted later

The creation of the first portable, electric-powered vacuum cleaner is usually credited to James Spangler, a part-time inventor who was working as a janitor in an Ohio department store. He realised that the dust raised by the Bissell carpet sweeper he used on the job aggravated his asthmatic cough. Inspired by a street cleaning machine, Spangler designed and made a prototype vacuum cleaner. It consisted of an electric fan motor driving home-made fan blades housed in a soap box with a cylindrical sweeping brush at the front and a pillow case as a dust collector, all pushed along by a broom handle. Spangler improved his prototype, for example by incorporating a shaft-driven front roller brush, and received a patent for his invention in 1908 (Figure 5.3a). Spangler then formed the Electric Suction Sweeper Company to manufacture his cleaner. One of his first customers was his cousin, who was married to William Hoover, a saddle-maker and leather merchant looking for a new business. Hoover was so impressed with his wife's vacuum cleaner that he bought into Spangler's business and patents. Hoover had Spangler's machine redesigned to produce the 'Model O', the world's first commercially successful portable electric vacuum cleaner (Figure 5.3b). Over the years Hoover's company made many improvements to the Model 0 in a series of new designs (Figure 5.4 middle and left). An important innovation was adding beater bars to the rotating front brush to loosen the dust in the carpet before it was sucked up, allowing Hoover its famous advertising slogan 'it beats as it sweeps as it cleans'. At $60 to $110 (about £1,000 to £1,700 today) a vacuum cleaner was a major purchase, so to sell his machines Hoover employed door-to-door salesmen to give home demonstrations and offered a 10-day, free home trial. By the 1920s several American companies were making vacuum cleaners, which were sold mainly to wealthy consumers (Gantz, 2012).

A second type of vacuum cleaner, the tank or canister design, which had an electric motor and fan mounted above a dust container and simply sucked up the dust via a hose, eliminated the rotating brushes and beaters of upright machines. One of the early tank cleaners, and the first European electric vacuum cleaner, was the Danish Nilfisk C1 of 1910. By 1921 the Swedish firm Electrolux had introduced the Model V, the first cleaner with a horizontal cylindrical tank mounted on runners to make it easy to move around (Figure 5.5). This design classic sold very well in Europe and the US.

Dominant designs

By the early 1920s the two dominant designs of vacuum cleaner – upright and tank (cylinder or canister) forms, which survive to the present – had been established. The 1920s therefore saw the transition of the vacuum cleaner industry from the invention and technical development phase to a large-scale production and sales phase (Gantz, 2012). Uprights had a fan driven by the motor through which the dirty air passed before being blown into the dust-bag – a 'dirty' air system. In tank cleaners the dirty air was sucked directly into a dust-bag or container and filtered by the bag or a filter before passing through the fan – a 'clean' air system.

From the 1920s to about 1950 vacuum cleaners were still expensive products mainly aimed at wealthy households. To cater for this luxury market, European manufacturers such as Electrolux started involving architects and industrial designers in designing their products, which therefore looked modern and sleek

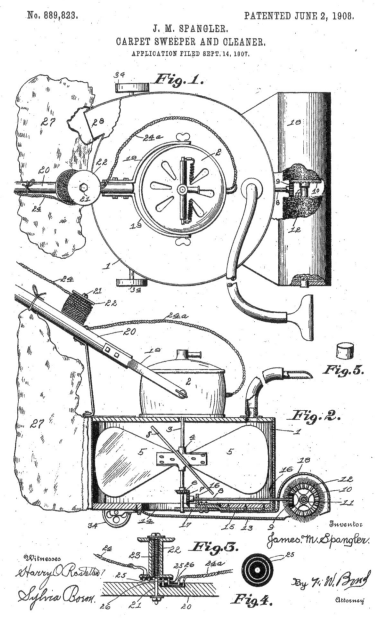

FIGURE 5.3a   Spanger's vacuum cleaner patent, 1908

FIGURE 5.3b Hoover (Electric Suction Sweeper Company) Model 0, 1908

FIGURE 5.4 (left) British Magic Appliances vacuum cleaner; (middle and right) Hoover upright cleaners, 1919 model and Hoover Junior 1936

compared to most American products designed by engineers. So, when demand for their appliances declined during the Great Depression of the 1930s, American manufacturers also started to employ industrial designers in an attempt to boost sales and reduce costs. Hoover, for example, retained a famous industrial designer, Henry Dreyfuss, to modernise its products. One of Dreyfuss' iconic designs was the Hoover 150 of 1936. It had a teardrop-shaped Bakelite motor cover incorporating the company's earlier innovation of a headlight, plus several other improvements such as a sculptured, lightweight magnesium body, a height adjuster, bag-full indicator and the ability to plug cleaning tools into the machine with the motor running. However, at $80 (about $1,400 or £900 today) the 150 was aimed at wealthy consumers and Hoover continued to make traditional-looking machines for the lower price market.

Another iconic design was the 1937 Electrolux Model XXX cylinder cleaner styled by an American industrial designer, Lurelle Guild. It had aluminium streamlines along its casing to suggest speed and was so successful that it remained in production until 1954 (Figure 5.6).

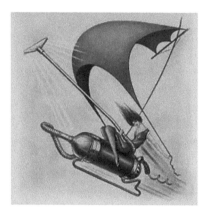

FIGURE 5.5 Electrolux Model V cylinder vacuum cleaner, 1921

FIGURE 5.6 Electrolux Model XXX (30) cylinder cleaner, 1937, styled by American industrial designer, Lurelle Guild

Gantz (2012) writes,

> For these early industrial designers, getting into the business was like shooting sitting ducks. Every product ... was ugly, ungainly and obsolete in style having been designed by engineers who were totally focused on functional performance, but oblivious to the new modern design trends and unaware of the public desire for more attractive appearance.

Production of vacuum cleaners almost ceased during the Second World War. But by the 1950s vacuum cleaners were becoming commonplace household equipment. Tank cleaners were the most common type in most countries, apart from Britain and the US where uprights were popular.

In this dominant design period, which lasted from the 1930s to the 1990s, manufacturers regularly introduced new models with incremental engineering and usability improvements and in updated styles and bright colours (Figure 5.7a). The changes included an increasing use of disposable bags, followed later by HEPA (high efficiency particulate air) bags and filters to remove fine dust and allergens such as dust mite excretions and pollen from the cleaner's exhaust air. Many uprights were redesigned to use a clean air system, like tank cleaners. Tank cleaners with an air turbine or motor-driven brushes in their cleaning heads were developed to improve their carpet cleaning performance and uprights began to have cleaning tools with long hoses for use on stairs and furnishings. Many of the improvements aimed at making vacuum cleaners more convenient to use. In 1957 Hoover introduced a Convertible upright, which automatically boosted power when the cleaning tool hose was plugged in. By the 1970s many cleaners had retractable electrical cords and a few models had power drive. Styling changes included uprights with a plastic case enclosing the dust bag, giving them a more contemporary look (see Figure 5.8b).

For frequent, light cleaning jobs a different type of vacuum cleaner, the hand-held 'stick' design was developed. Stick cleaners were first seen in products such as the 1945 Regina Electrikbroom, but were produced by several manufacturers from the 1960s. This was to meet changing cleaning habits and also to persuade consumers to buy a second vacuum cleaner.

Important changes were also made in the use of materials and in manufacturing to reduce costs. For example, from the 1960s manufacturers replaced steel and cast aluminium parts with injection-moulded high-impact plastics, with the result that the products were lighter as well as cheaper to make. Vacuum cleaners could then be economically mass-produced by moulding sets of plastic components, making and buying in other components and sub-assemblies such as motors and brushes, and putting together the various components and sub-assemblies on an assembly line.

The first *Which?* report on vacuum cleaners was in 1960. It identified over 100 models on the British market of which 21 were tested: four uprights, 16 tanks (cylinder, canister or spherical shaped), and one hand-held stick design.

Vacuum cleaners **105**

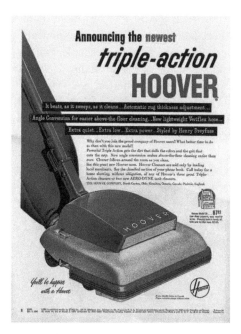

FIGURE 5.7a   Advertisement for Hoover Model 29 in bright red, 1950, 'styled by Henry Dreyfuss'

FIGURE 5.7b   Hoover Constellation 1958, which floated on a cushion of its own exhaust air and remained in production until 1980. The distinctive spherical design (again by Henry Dreyfuss) reflected late 1950s American concerns with space following the launch of the Russian Sputnik 1 satellite in 1957

106 Vacuum cleaners

The cleaners tested cost from £7.50 to £40 (about £150 to £800 today) and only differed in materials, some mechanical components and aesthetic design from vacuums of the 1920s and 1930s. For example, two classic designs: the 1959 upright Hoover DeLuxe/Senior had a streamlined plastic motor cover with a built-in headlight, while the 1958 spherical tank Hoover Constellation floated like a hovercraft on a cushion of its exhaust air (Figure 5.7b). *Which?* found that the uprights with their revolving brushes and beaters were more efficient than most tanks at removing a sample of dust from a standard length of carpet. But one tank with beaters in its cleaning head was almost as efficient as the uprights. The stick design had average cleaning performance, but was very convenient to use. About half of the vacuums used a paper dust bag (often as an option), which was

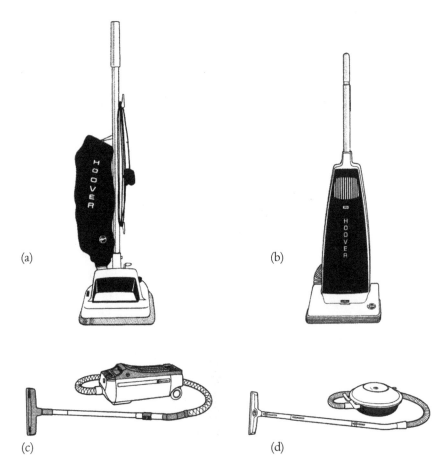

FIGURE 5.8 Typical late 1970s vacuum cleaners on the British market:
(a) Hoover Dirtsearcher Junior U1016
(b) Hoover Convertible U5010
(c) Electrolux Automatic 345
(d) Hoover Celebrity S3005, an updated version of the floating Constellation

convenient to dispose of, but reduced the suction of some cleaners by up to 40% as it filled (Consumers' Association, 1960).

Although the market remained dominated by tank (mainly cylinder shaped) and upright cleaners – Figures 5.8a to 5.8d (Consumers' Association, 1977, 1978) – two new types of vacuum cleaner were introduced in an attempt by manufacturers to meet new consumer needs and increase sales in the saturated markets of the 1980s and 1990s. These were multi-purpose 'wet and dry' tank cleaners for sucking up liquids or workshop and garden debris and small, hand-held mains or battery powered vacuums for minor cleaning and dusting jobs, pioneered by Black and Decker's Dustbuster (Figure 5.9). These new types of cleaner helped boost sales as they were usually bought in addition to a standard machine.

Disruptive innovation

While existing vacuum cleaner manufacturers continued to make incremental improvements and styling changes to their products, a radically innovative type of vacuum cleaner was created by a British inventor and designer, James Dyson. Dyson's cyclonic cleaner, which eliminated the need for a bag and so did not lose suction as it filled with dust, proved to be a disruptive innovation. Its great commercial success resulted in most of the established manufacturers – several of whom had rejected Dyson's original offer to license his invention – developing their own bagless and cyclonic cleaners. Consequently, Dyson was forced to defend his patents against infringement.

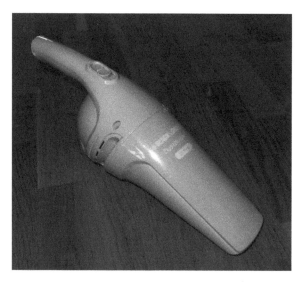

FIGURE 5.9 Recent model of the innovative hand-held, rechargeable battery-powered Black and Decker Dustbuster, first launched in the USA in 1979

CASE STUDY: THE DYSON CYCLONIC CLEANER

The cyclone project began in 1978 when Dyson had been using a Hoover upright vacuum cleaner and, annoyed by its inefficiency, replaced it with a powerful cylinder cleaner. He was also struck by the deficiencies of the cylinder cleaner, in particular because its suction rapidly declined when its collection bag became clogged by dust before it was full. Dyson's creative breakthrough came with his idea of *transferring* industrial cyclone technology to a vacuum cleaner. (Although Dyson was not aware of it at the time, a bagless vacuum cleaner that used centrifugal force and a water filter to separate dust from dirty air had been developed and launched in 1937 by Rexair, an American manufacturer. Luckily Rexair's patents were not similar enough to Dyson's cyclone idea to prevent him from patenting his invention (Gantz, 2012)).

Dyson was an entrepreneur as well as an inventor–designer and one of his products was a novel wheelbarrow, the Ballbarrow, which used a ball instead of a front wheel. In order to extract paint dust from the powder coating plant in the Ballbarrow factory, an industrial cyclone had been installed. In such a device a rotating air flow is created within a cylindrical or conical container called a cyclone. The air rotates so fast that any particles in the air are forced to the walls of the cyclone from where they spiral down and can be collected, whilst clean air moves to the centre. Dyson had the idea that a miniature version might be applied to a vacuum cleaner. He tested his idea by fitting a cardboard cyclone to his old Hoover upright and found that the cyclone principle did work.

It took another four years, which involved him making and testing over 5,000 prototype cyclones (Figure 5.10a right) and much tinkering with mock-ups and prototypes, before Dyson was ready to show his cleaner to potential licensees. He could not patent the application of a single cyclone, since that would only improve an existing technology. So he developed a dual cyclone system, which used the first to separate larger items of dirt and the second to extract finer dust particles. In 1980 Dyson applied for a patent for a dual cyclone vacuum cleaner. By 1982 he had embodied his invention in the design of a prototype dual cyclone cleaner, which eliminated the need for a dust bag, to demonstrate to potential manufacturers (Figure 5.10a centre). However, the established appliance manufacturers proved extremely reluctant to license Dyson's invention. In a classic case of the *'not-invented-here'* syndrome they remained wedded to existing vacuum cleaner designs and were unconvinced by Dyson's cyclonic technology (Coren with Dyson, 2001).

Dyson continued to develop his cyclonic cleaner, using both his engineering and industrial design skills, and eventually succeeded in licensing a new design in Japan. The Japanese-made 'G-Force' cleaner was an unusual colour – pink – and looked quite different from conventional upright vacuum cleaners

(Figures 5.10a left and 5.10b). Following its launch in 1986 the G-Force sold successfully to Japanese design-conscious consumers for the equivalent of £1,200. Thus Dyson's invention had, through design, finally become a commercial product innovation (Coran with Dyson, 2001).

The royalties from sales of the G-Force, plus other products he had developed, enabled Dyson to start his own business to design and make cyclonic vacuum cleaners. He assembled a small team of design engineers and began the task of developing a bagless cyclonic upright cleaner aimed at the mass market. Dyson's team put great effort into considering how people would use the product, as well as making it function as efficiently as possible, look attractive and be economical to manufacture. For example, its handle doubled as a telescopic hose and tool for cleaning stairs, etc., the transparent outer cyclone allowed the user to see the dust being collected, it was coloured yellow and grey, and the design made extensive use of moulded plastics. The first Dyson Dual Cyclone cleaner, the DC01 (Figure 5.11a), was made and launched in the UK in 1993. This was 15 years after Dyson's initial creative idea, demonstrating how long it often takes to get ideas and inventions into mass production (Open University, 2014).

Which? tested the DC01 in 1994. The magazine reported that, unlike vacuum cleaners with bags, the DC01 met its claims of not losing suction as it filled and produced little exhaust dust. However, conventional uprights picked

FIGURE 5.10a (left) Dyson's first production cleaner with concentric cyclones, the G-Force, made in Japan; (centre) the first prototype in which the cyclones were placed side by side; (right) a few of the models used to develop the best shape of cyclone

FIGURE 5.10b Dyson's pink and lavender G-Force dual cyclone vacuum cleaner, 1986, made and sold in Japan

up more dust when empty and the Dyson was heavy, noisy and expensive; so although it performed well the DC01 did not rate as a Best Buy (Consumers' Association, 1994). Despite selling for £200 (about double the price of most other vacuums), Dyson's innovation proved to be very successful and quickly gained nearly a fifth of the British market. A cylinder version, the DC02 (Figure 5.11b) and an improved version of the DC01 followed, both of which performed well in *Which?* tests, resulting in the DC01 becoming a Best Buy (Which? Ltd., 1996a, 1996b).

Following these early models, Dyson's company has invested heavily in the research, design and development of a series of bagless cyclonic cleaners. Significant innovations and improvements have been made to the cleaners' cyclone and electric motor technologies, their usability and to address their initially poor reliability. The products now include ranges of upright and cylinder cleaners that have multiple 'root cyclones' plus washable filters to remove very fine particles. Some models have a ball instead of wheels to provide better manoeuvrability (Figure 5.12) and cyclones with oscillating nozzles, to collect even the finest particles without filters (Figure 5.13a). Dyson's product development process thus represents an outstanding example of competing through continuous innovation and design. Since 2002, to reduce costs and the bring component suppliers closer to the factory, Dyson cleaners have been manufactured in Malaysia for export all over the world.

Vacuum cleaners 111

FIGURE 5.11a Dyson DC01 Dual Cyclone bagless upright cleaner, made and launched in the UK in 1993

FIGURE 5.11b Dyson DC02 bagless cyclonic cylinder cleaner, launched in 1995

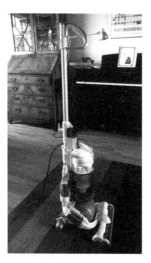

FIGURE 5.12 Dyson DC24 multi-floor lightweight upright cleaner with several root cyclones and highly manoeuvrable Dyson Ball™ steering, 2012

The impact of Dyson's innovation

The success of the Dyson bagless cleaners, which by 1991 had captured half of the UK market by value, stimulated other vacuum cleaner manufacturers to develop new products and especially their own bagless designs. The disruptive effect of Dyson's innovation thus set off a new divergent phase of vacuum cleaner design. As *Which?* commented,

> it was undoubtedly the launch of Dyson's bagless ... cleaner in 1993 that injected new life into the market. It set new standards for continuous dust pickup ... such technological change combined with Dyson's fresh approach to design has prompted many rival manufacturers to reassess their own products.
>
> (Which? Ltd., 2001a)

Dyson had the patents on cleaners with multiple cyclones, so rival manufacturers should have been prevented from introducing products that infringed his patents. However, this did not stop some companies trying to copy Dyson's innovation. Hoover, for example, lost a court case against Dyson in 2000 when it launched a bagless cleaner with three cyclonic stages. Since then Dyson's competitors have adopted various approaches to designing bagless cleaners. They include either using filters (instead of a bag) to catch the dust before it leaves a collecting container, or using a single cyclone plus filters for fine dust. As Dyson's original patents expired from 2001, many other manufacturers have launched their own multi-cyclonic cleaners, such as Hoover's Airvolution range. Dyson now has to compete through investment in consumer and technical research, continuous patenting, engineering development and innovative design. Dyson's 'form follows function' design philosophy has resulted in increasingly functional-looking products (e.g. Figure 5.13a). Other manufacturers have tended towards a more conventional styling approach, with products that are increasingly sleek and sculptural in form (e.g. Figure 5.13b).

As well as a variety of bagless and cyclonic cleaners, there have been other innovations resulting from the upsurge of interest in vacuum cleaner design and technology triggered by Dyson's products. The most significant of these innovations are robotic and cordless vacuum cleaners.

Robot vacuum cleaners

The idea of a robot vacuum cleaner is not new. In 1957 RCA-Whirlpool demonstrated a robot vacuum cleaner (actually it was remote-controlled) in a mock-up of its automated Miracle Kitchen. However, it wasn't until 2001 that Dyson designed and trialled one of the first robot vacuum cleaners, but he did not put it into production because of its weight, complexity and cost. A year later the first production robot cleaners – the Trilobite from Electrolux and the Roomba from iRobot, a US robotics company – were launched. *Which?* tested both products

FIGURE 5.13a Form follows function: Dyson DC75 Cinetic Big Ball with multiple oscillating tip cyclones to remove the finest particles without a filter, 2015

FIGURE 5.13b Function plus sleek styling: Miele Dynamic U1 Powerline upright bagged vacuum cleaner. Its 1500W motor meets the requirements of the EU Energy Label, 2014

and concluded that, although they could navigate a room, neither could replace a conventional vacuum cleaner because of their relatively small capacity, poor dust pick-up and ability to clean into corners (Which? Ltd., 2001b, 2003). Since then several manufacturers have introduced robot cleaners; some were really carpet sweepers while others had a vacuum function. *Which?* has found that these new robot vacuums are convenient and many now perform well. However, for most consumers robot cleaners are unlikely to replace a conventional vacuum as, so far, they are relatively slow, cannot clean stairs or upholstery and are expensive (Which? Ltd., 2015).

Cordless vacuum cleaners

The first battery-powered vacuum cleaners were small hand-held products for collecting crumbs and spills, such as the 1979 Dustbuster (Figure 5.9). Since then there has been a growing market for cordless vacuum cleaners that are lighter and more convenient to store and use than a mains-powered cleaner, but are suitable for whole-house cleaning. Improvements in rechargeable batteries – a transfer of technology – have enabled manufacturers to develop better cordless vacuum cleaners. *Which?* tests showed that most cordless models did not perform as well as a conventional cleaner. However, a few cordless stick and upright designs already had a good enough cleaning performance and battery life to be the only vacuum cleaner needed in the home (Which? Ltd., 2014).

Environmental impacts and regulation

Concern about the environmental impacts of vacuum cleaners occurred later than for most other consumer electrical products. The first mention of vacuum cleaner running costs by *Which?* occurred in 1981 when the magazine noted that most upright cleaners had 400 watt motors, which if used for three hours a week would cost about £3 per year in electricity (Consumers' Association, 1981). Since then vacuum cleaner motors have become more powerful, ranging from 500 to 2400 watts. But it was not until 2003 that *Which?* vacuum cleaner tests took energy use and noise into account, with its Best Buys typically having 1200–1600-watt motors and tank types generally being quieter than uprights.

A number of life cycle assessment (LCA) studies of vacuum cleaners have been carried out. These have shown, in common with other electricity-using consumer products, that their dominant impacts arise during use of the product, with only relatively minor impacts arising from manufacture, distribution and disposal or recycling. For example, one study showed that 90% of the environmental impacts (on natural ecosystems, resource use and human health) of an upright with an 800-watt, and a cylinder with a 1300-watt motor occurred at the use phase (Kota and Chakrabarti, 2007). Dyson's company conducted a LCA of its DC50 upright cleaner and reported that 91% of the carbon emissions during the cleaner's life occurred during consumer use (Dyson, 2015). Another study, which took into account materials processing and resources, indicated that important impacts can arise from the depletion of non-renewable materials used in a vacuum cleaner, such as copper for the motor. For cleaners that used dust bags, their production and disposal had only minor impacts (Hur *et al.*, 2005).

The view that vacuum cleaners generate relatively minor environmental impacts continued until 2013 when the EU announced that it would introduce a vacuum cleaner Energy Label. From September 2014, every vacuum cleaner offered for sale in the EU has an Energy Label and is rated from A–G for energy use, dust pick-up, dust re-emission and noise. In addition to saving energy, under the EU's Ecodesign of Energy-related Products Directive no EU cleaner should have a motor more powerful than 1600 watts. This means that in average use a cleaner should use no more than 62 kWh per year (similar to that of an energy-efficient television). A cleaner rated 'A' on the Energy Label should use under 28 kWh per year, pick up 90–100% of dust from carpets and floors and re-emit only 0.02% of the collected dust. From 2017 the maximum power will be reduced to 900 watts with 43 KWh per year maximum average energy use (Top Ten EU, 2013).

Another neglected aspect of the environmental impact of vacuum cleaners is their durability. In the past many vacuum cleaners had a very long life and hence were rarely discarded and replaced. Modern cleaners tend to be less reliable and do not usually last as long. However, *Which?* surveys have found that 80–90% of several vacuum cleaner brands worked for six years without repair and so should last many years before being discarded or recycled (Which? Ltd., 2004). Nevertheless, a major survey conducted in the 1990s found that, instead of having them repaired, on average British consumers threw away their vacuum cleaners after only eight years (Cooper, 2004).

Social influences and impacts

Vacuum cleaner technology and design and their adoption and use have long been influenced by socio-economic and cultural factors.

As noted earlier, between the two World Wars vacuum cleaners were luxury products mainly bought by the upper middle classes as way of reducing their need for resident domestic servants. Forty (1986) describes how manufacturers and advertisers constantly promoted the idea that domestic appliances were labour-saving devices that could substitute for servants. This was despite the fact that those who could afford a vacuum cleaner could also afford to employ domestic help. At the same time, as was noted in the washing machine chapter, middle-class concern with cleanliness had increased since Victorian times. This allowed manufacturers also to promote vacuum cleaners as the most effective way of removing unhygienic dust, dirt and germs. Social historians of technology and design such as Cowan (1983) and Forty (1986) have therefore argued that domestic appliances *increased* rather than saved women's labour by persuading them of the need for, and enabling ever-higher standards of, cleanliness and hygiene. Labour-saving appliances such as vacuum cleaners were also symbols of modernity in the home, valued for their symbolic status as much as for their function and efficiency (Ryan, 2006).

Designers of vacuum cleaners also responded in various ways to social changes and concerns. For example, before the Second World War US vacuum cleaner manufacturers became caught in 'sanitation fever' and some responded with designs that contained chemicals to sterilise germs (Gantz, 2012, p. 102). By the 1950s – as vacuum cleaners became an increasingly commonplace domestic appliance and when women who formerly had home help were doing their own housework – manufacturers focused on designing more powerful products that were faster and easier to use. From the 1970s, as more women worked outside the home and some men started doing the housework, convenience became more important than regular deep cleaning, thus stimulating the development of hand-held, stick and cordless vacuum cleaners. A new upsurge in concern about health, in particular dust, allergens and pets causing problems for asthma sufferers, has led since the 1990s to the introduction of bags, filters and cyclonic methods for removing finer and finer particles from vacuum cleaners' exhaust air.

Future developments

Vacuum cleaners are one of the few domestic products that still require their users to do physical work – carrying, operating, emptying and storing. That is why much innovation in their technology and design now focuses on attempts to automate this chore with robot vacuum cleaners, including, for example, robots that can climb and clean stairs. Meanwhile, Dyson has developed the 360 Eye™ robot vacuum cleaner, launched in 2015 (Figure 5.14). A panoramic camera and infrared sensors allow the cleaner to visualise the room, work out its location and use landmarks in the room to navigate its way around. The cleaner can be controlled via a mobile app which allows users to set it to clean at specific times (Robarts, 2014).

Since Dyson's ranges of cyclonic cleaners and robot machines, there have been hardly any other significant innovations in vacuum cleaner technology. However, one new cleaning technology from a British inventor is called Air Recycling Technology. This uses a pressurised air stream, rather like a combined garden and house vacuum cleaner, to both blow and suck dirt into a dust separator. Working prototypes have been shown to be more energy-efficient than conventional vacuums, but so far the technology is not currently used in any production cleaner (Edginton, 2005).

FIGURE 5.14 Dyson 360 Eye™ robot vacuum cleaner, 2015

The pattern of innovation

Innovation in vacuum cleaners has followed the familiar pattern shown by several other consumer products discussed in this book. It started with an early divergent experimental phase, followed by the establishment of two dominant designs, which were steadily improved in design and manufacturing methods. Then, after the disruptive innovation of Dyson's cyclonic cleaner, an innovative design phase emerged with many designs of bagless, cyclonic, hand-held, cordless and robot products developed and introduced by existing manufacturers as well as new entrants to the industry.

References

Bellis, M. (2015) Invention and History of Vacuum Cleaners [Online]. Available at http://inventors.about.com/od/uvstartinventions/a/Vacuum-Cleaners.htm (accessed January 2015).
Bowden, S. and Offer, A. (1994) 'Household appliances and the use of time: the United States and Britain since the 1920s', *Economic History Review*, vol. XLVII, no. 4, pp. 725–748.
Consumers' Association (1960) 'Vacuum cleaners', *Which?* April, pp. 72–82.
Consumers' Association (1977) 'Upright vacuum cleaners', *Which?* January, p. 6.
Consumers' Association (1978) 'Cylinder vacuum cleaners', *Which?* June, p. 362.
Consumers' Association (1981) 'Upright vacuum cleaners', *Which?* April, pp. 210–215.
Consumers' Association (1994) 'Cleaning up', *Which?* April, pp. 52–56.
Cooper, T. (2004) 'Inadequate life? Evidence of consumer attitudes to product obsolescence', *Journal of Consumer Policy*, vol. 27, pp. 421–449.
Coren, G. with Dyson, J. (2001) *Against the Odds. An autobiography* (2nd edn), London, Texere.
Cowan, R.S. (1983) *More Work for Mother: The Ironies of Household Technology from the Open Hearth to the Microwave*, New York, Basic Books.
Dyson (2015) Engineered for efficiency [Online]. Available at http://www.dyson.co.uk/community/corporate-social-responsibility.aspx (accessed January 2015).
Edginton, B. (2005) The air recycling cleaner [Online]. Available at http://www.g0cwt.co.uk/arc/index.htm (accessed January 2015).
Floating Path (2013) Global penetration rates in the home appliance market, 17 July [Online]. Available at http://www.floatingpath.com/2013/07/17/global-penetration-rates-of-home-appliance-ownership/ (accessed January 2015).
Forty, A. (1986) *Objects of Desire. Design and Society 1750–1980*, London, Thames and Hudson.
Gantz, C. (2012*) The Vacuum Cleaner: A History*, Jefferson, North Carolina and London, MacFarland and Co.
Hur, T., Lee, J., Ryu, J. and Kwon E. (2005) 'Simplified LCA and matrix methods in identifying the environmental aspects of a product system', *Journal of Environmental Management,* vol. 75, pp. 229–237.
Kota, S. and Chakrabarti, A. (2007) 'Use of DFE methodologies and tools – major barriers and challenges', *ICED '07 International Conference on Engineering Design*, 28–31 August 2007, Cité Des Sciences Et De L'industrie, Paris, France.

Open University (2014) 'Exploring innovation', Block 1, T317 *Innovation designing for change*, Milton Keynes, The Open University.

Robarts, S. (2014) Dyson 360 Eye robotic vacuum cleaner 'sees' its environment [Online], *Gizmag*, 5 September. Available at http://www.gizmag.com/dyson-36-eye-robot-vacuum-cleaner/33686/ (accessed February 2015).

Ryan, D.S. (2006) 'The vacuum cleaner under the stairs: women, modernity and domestic technology in Britain between the wars', *Design and Evolution: The Proceedings of the Design History Society annual conference*, Delft, Netherlands, Design History Society and Technical University, Delft.

Top Ten EU (2013) Vacuum cleaners: Recommendations for policy design, Paris, Top Ten International Group, August [Online]. Available at http://www.topten.eu/uploads/File/Vacuum%20cleaners_Policy%20Recommendations_Aug_13.pdf (accessed January 2015).

van Dulken, S. (2002) *Inventing the 20th Century*, London, British Library.

Which? Ltd. (1996a) 'Vacuum cleaners', *Which?* April, pp. 38–43.

Which? Ltd. (1996b) 'Best Buy update', *Which?* December, p. 44.

Which? Ltd. (2001a) 'Design of the times', *Which?* April, pp. 51–54.

Which? Ltd. (2001b) 'First robotic vac set loose', *Which?* March, p. 14.

Which? Ltd. (2003) 'Roomba service', *Which?* December, p. 49.

Which? Ltd. (2004) 'Brand reliability', *Which?* May, p. 12.

Which? Ltd. (2014) 'Corded vacs bite the dust?' *Which?* September, pp. 54–57.

Which? Ltd. (2015) 'Robot vacuum cleaners: first look reviews', *Which?* [Online]. Available at http://www.which.co.uk/home-and-garden/home-appliances/reviews-ns/robot-vacuum-cleaner-first-look-reviews/ (accessed January 2015).

6
MOBILE PHONES

Along with television (Chapter 4), the mobile phone is one of the best examples of a product which has not only undergone radical changes in its technology and design but has had major cultural, socio-cultural, political and environmental impacts. The mobile (or cell) phone has evolved from the invention by US Bell Labs of the cellular radio network for 'car phones' aimed at executives and celebrities, to a global mass market industry. One result is that by 2014 nearly 2 billion mobile phones, of which 1.3 billion were smartphones, were sold worldwide and in the industrialised world most people have a mobile phone (IDC, 2014). In the UK, for example, in 2014, 93% of adults had a mobile phone and 61% of adults, rising to 88% of those aged 16-24, had a smartphone (Ofcom, 2014). A smartphone is essentially a powerful hand-held two-way radio and computer offering not just phone calls and text messaging but Internet access, email, music player, camera, games, GPS mapping, online radio, TV and films, and many other applications ('apps') in a pocket or handbag.

In developing and newly industrialised countries, mobile phones are being widely adopted because of the lack of investment in an infrastructure of fixed line telephones. For example, in 2010 three-quarters of phone calls in China and nearly all (98%) calls in India were via mobiles (Ofcom, 2011a – Figure 6.1). By 2013, 88% of China's population had a smartphone (Ofcom, 2013a). The phones concerned may be used ones returned for recycling by consumers in industrialised countries when they upgrade to the latest model.

Mobile phone technology

The modern mobile phone is the result of over 60 years of innovation in electronics and product design through (so far) five generations of telecommunications technology involving many different systems and standards. Only the main developments are covered in the brief review below.

120 Mobile phones

FIGURE 6.1 Most telephone communications in developing and newly industrialised countries, such as India, are by mobile phone

Zero generation (0G) technology

The first commercial mobile phone call was made in the US as long ago as 1946 from a car phone fitted with a radio transmitter and receiver big enough to fill the car's boot (trunk). It used MTS (mobile telephone system) technology developed during the Second World War by Bell Labs and AT&T. Like early telephones, the system required users to contact an operator to make a call, press a button when they wanted to speak and release it in order to listen. Within a city served by the MTS only up to three subscribers could make calls at the same time, coverage was only about 20–25 miles (32–40 km) and the equipment and calls were very costly. Not surprisingly, this early mobile phone system had a very limited market.

The next milestone in mobile car phone technology was IMTS (Improved Mobile Telephone Service), released in 1964 by Bell System. IMTS allowed direct dialling without an operator, but was limited to single-user frequencies, which made it incapable of handling lots of users, due to a lack of frequencies available for mobile phones on the radio spectrum.

This limitation had been solved – at least in theory – by D.H. Ring and W.R. Young who in December 1947 published a Bell Laboratories' article entitled 'Mobile Telephony – Wide Area Coverage'. They proposed the concept of a cellular network in which an area is divided up into small regions called cells, containing radio base stations. This allowed limited available radio frequencies to be reused again and again by different users, provided that no two adjacent cells used the same frequency. It meant that as a person moved within the network their telephone call would be passed from one cell to another, automatically switching to the new frequency. When the MTS/IMTS systems were introduced, the technology to implement these ideas did not exist, nor had the frequencies been allocated. The subsequent development of cellular technology and the system of frequency reuse by Bell Labs' engineers and others in the 1960s and 1970s overcame the radio spectrum limitations of the old car telephone service, allowing millions of users to make calls at once and heralding the arrival of the modern mobile phone.

First generation (1G) analogue technology

However, it was not until 1974, when the Federal Communications Commission (FCC) in the US released part of the radio frequency spectrum, that the first experimental cellular communication could take place. In 1977 Illinois Bell installed the first cellular telephone network in Chicago with just 10 cells. The successful operation of this network led to the development of the first fully functional US analogue cellular system, the Advanced Mobile Phone System (AMPS) in 1983. Also in 1983 the mobile phone prototyped by Motorola during the 1970s was launched as the world's first commercial hand-held cell phone, the Motorola DynaTAC 8000X, priced at nearly $4,000 (see Figure 6.3).

In Europe, mobile phone development was led by the Scandinavian countries, collectively creating the Nordic Mobile Telephony (NMT) analogue cellular system, introduced in Sweden and Norway in 1981 and Denmark and Finland in 1982. But at first these cellular networks served car radio phones and bulky 'transportable' cell phones (Figure 6.2) before hand-held designs became available.

In the UK, government licences to provide mobile phone services were awarded to British Telecom in partnership with Securicor, leading to the creation of Cellnet and to a joint venture company, Vodafone (VOice and DAta over the teleFONE), which won a bid for the second licence. Both of these networks adopted a version of the AMPS analogue system, called TACS (Total Access Communication System) and launched their services in January 1985. Other variants of AMPS technology called ETACS and JTACS were used in the 1980s and 1990s in Europe, Japan and elsewhere; but almost all have been discontinued since the development of digital mobile phone systems (University of Salford, 2013; Wikipedia, 2013a; Retrobrick, 2013).

Second generation (2G) digital technology

In the 1990s, 'second generation' (2G) mobile phone systems emerged. Two systems competed: the European GSM (Global System for Mobile communications) and the American CDMA (Code Division Multiple Access) but, except in the US, GSM became dominant worldwide. 2G differed from 1G by using digital instead of analogue transmission. In 1991 the first GSM mobile phone network was launched in Finland, and the UK's first GSM network went live in 1992. Digital technology also enabled new services such as SMS (short message service) text messaging, first introduced in Finland in 1993. The advent of prepaid services in the late 1990s soon made 'texting' the communication method of choice amongst the young, which quickly spread across all age groups. 2G digital technology also enabled mobile phones to access the Internet, but only with an early system called WAP (wireless application protocol) that provided limited services. However, manufacturers soon realised that the mobile had enormous potential for data communications. Consequently, the mobile phone started to move away from being a device that could only make telephone calls and began to take on more computer-like functions, leading to the emergence of the smartphone. With fashionable new designs and additional functions, by 1999 UK mobile phone ownership had doubled to reach nearly half (46%) of the population and increased to over three-quarters (76%) by 2000 (University of Salford, 2013; Wikipedia, 2013a).

Anticipating this massive growth, a 1994 report in *Which?* said 'No longer the ultimate yuppie toy, the mobile phone has gone mass market.' The same report explained in lay terms how mobile phones work:

> When you make a call from a mobile phone, it transmits radio waves either in analogue form (waves) or digital (computer code) form. If the phone is within range these signals are picked up by a base station owned by the various cellular networks. If you have a contract with the network then it will recognise your phone. The network does this through the information held on the phone's 'smart card' [subscriber identity module or SIM card] for digital phones or by the phone number and the phone's electronic serial number which is programmed into it.
>
> *(Consumers' Association, 1994)*

Third generation (3G) technology

As the use of 2G phones became more widespread, it became clear that demand for data, such the ability to browse the Internet and download music, was growing. 2G technology could not provide the data transmission rates required, so the industry began to work on the next generation of technology known as 3G. The main technological difference that distinguishes 3G from 2G is the use of packet switching rather than circuit switching for data transmission. (Circuit-switched networks reserve a dedicated channel for the entire communication. In packet-switched networks the message gets broken into small data packets, each with an address that seeks the most efficient route as circuits become available and describes the sequence for reassembly at the destination, thus sharing the network's capacity between multiple communications.) As with 2G, there were rival 3G standards used in different countries and by different providers. In Japan, NTT DoCoMo launched the first commercial 3G network in October 2001. A year later there were European launches of 3G in Italy and the UK.

In the UK, 3G provided faster access to WAP, the limited mobile Internet accessed via each network's website rather than offering full Internet access. The higher connection speeds of 3G technology enabled for the first time video calls, streaming or downloading of audio and video content from WAP and other sites to 3G handsets. However, a *Which?* report in May 2005 found that 3G coverage, reception and services still needed improving. Given these limitations, take-up of 3G in the UK was slower than expected; for example, in 2005 only 13% of Vodafone's customers were 3G subscribers (Which? Ltd., 2005a). But towards the end of that year *Which?* reported that 3G services had improved and sales were increasing, although 3G was 'still waiting for the "killer application" to spur growth' (Which? Ltd., 2005b).

By the end of 2007, there were 295 million subscribers on 3G networks worldwide, about 9% of the total, the rest still being on 2G.

3G really came into its own with the development and marketing of increasingly sophisticated smartphones that combine phone, PDA (personal digital assistant) and other functions such as a word processor; one example being the Nokia 6680, a *Which?* Best Buy of 2005. Apple's touchscreen smartphone, the iPhone, when upgraded to a 3G smartphone version in 2008 was essentially a powerful hand-held computer that provided the kind of services (Figure 6.9 left) that could make best use of 3G technology, including fast Internet access, email, streaming and downloading music and video, GPS location and navigation, video calls, gaming and much more.

Fourth generation (4G) technology

Whereas 2G technology was suitable for making calls and sending text messages, and 3G made it possible to access the Internet effectively through a mobile phone, by 2009 it had become clear that 3G networks would sooner or later be overwhelmed by the growth of applications that required much higher data transmission rates. Consequently, the industry began developing 4th-generation technologies, with the promise of speed improvements up to tenfold over 3G. One of the main ways in which 4G differs from 3G technology is in employing an all-IP network for both data and voice transmission (IP means Internet Protocol – the set of rules governing how packets of information are transmitted over the Internet). The main benefit of 4G services is that they make it possible to browse the Internet on a mobile phone at speeds similar to those provided by home broadband. Because of this, 4G is ideally suited for services which demand more capacity like video streaming, route mapping and using social networking sites. For the typical UK user, download speeds of initial 4G networks should be around 5–7 times those for existing 3G networks. This means that a music album taking 20 minutes to download on a 3G phone at an average speed of 1Mbit/s takes just over 3 minutes on 4G at 6Mbit/s (Ofcom, 2013c).

Operating systems

In common with the developing telecommunication technologies, during their evolution mobile phones have used a number of different operating systems – the software built into the handset. The technical history is long and complex, but software has played a growing role in the phones' functionality. Powerful operating systems have become increasingly important, especially for smartphones. By 2013 there were four main operating systems (OS): Apple's iOS first introduced for the iPhone in 2007 and frequently updated, followed by Google's Android, also in many versions, Microsoft's Windows Phone and Blackberry's systems. According to *Which?* each OS has its pros and cons (Which? Ltd., 2013a). Apple's iOS is easy to use, integrates with other Apple products and has the biggest and best app store, which only supplies Apple-approved software. However, iOS is less customisable which makes it difficult for users to personalise their phones. In contrast Android, used in rivals to the iPhone such as the Samsung Galaxy range, is open source and highly customisable; for example, it allows the user to choose a different browser or touchscreen keyboard. But the Android apps store, while almost as big as Apple's, is unregulated with some poor quality apps and possibly a few apps that contain viruses. Windows Phone OS has a simple colourful interface and provides integration with Microsoft software and systems, but its apps store is relatively small and its OS is used by few manufacturers, notably Nokia whose mobile division was taken over by Microsoft in 2013. Blackberry's OS offers several unique features such as a homepage that shows all the apps running; Blackberry phones offer a touchscreen or a physical QWERTY keyboard, but the apps store is also small and the OS currently only runs on Blackberry phones.

Mobile phone design

The design of the mobile phone itself – the 'handset' – is a story of miniaturisation of the electronics and increasing functionality facilitated by the successive generations of telecommunications technology and operating systems. Over 30 years, many technological innovations enabled the design of the handset to evolve from the first heavy and expensive portable 'brick' cell phones, which only allowed users to make telephone calls, into the pocketable, multi-functional smartphone.

Before the introduction of the first iPhone in 2007 with a simple rectangular form and large touch-screen covering most of the phone's front face, which has become the dominant design, there were a variety of phone forms and hundreds of designs from different manufacturers. These included numerous 'candy bar' models with a small screen above a keypad; clamshell; slide; and horizontal format phones, and a few with a full QWERTY keyboard. Only a small selection of significant designs is shown here. But which designs are the most significant is not always agreed and several lists of 'iconic' handsets have been published, some celebrating 40 years since the first experimental mobile phone call in 1973 (e.g. Web Design, 2009; *Guardian*, 2010; *Guardian*, 2013; University of Salford, 2013; Skipworth,

2013; Baird, 2013). All the lists, however, include the Motorola DynaTAC 8000X 'brick' (Figure 6.3) as the first proper mobile and the iPhone (Figure 6.9) as the iconic smartphone.

Car radio phones

As mentioned earlier, before the development of hand-held models, 'mobile' phones were car radio telephones, first introduced for private commercial users in the US by Bell in 1946. The size of the VHF or UHF radio transmitters and receivers made these early car phones very large and heavy and hence non-portable. By 1964 the car phone service had improved; the user did not have to call an operator but still had to push a button to talk and listen. The equipment, although much smaller, was still too large and heavy to be carried except in a vehicle. The phones were dominated by heavy battery packs with a separate handset usually connected via a length of curly wire. Some later radio phones called 'transportables', such as Motorola's 4500x transportable phone (Figure 6.2), were designed to be carried both in and outside of a car, but were certainly not pocketable.

From 'bricks' to smartphones

Analogue phones

The first 'proper' personal hand-held mobile phone, the DynaTAC 8000X was launched in the US in 1983 following a decade of research, design and

FIGURE 6.2 Motorola 4500x 'transportable' car phone, 1988

development by Motorola in parallel with the development of cellular technology and the first cellular networks. The phone was the result of the vision of Dr. Martin Cooper, considered the inventor of the first portable handset, and the first person to make a call on a prototype portable cell phone in 1973. One of the original Motorola design team members responsible for creating the DynaTAC 8000X recalled, 'Marty (Cooper) called me to his office one day in December 1972 and said, "We've got to build a portable cell phone" and I said, "What the hell's a portable cell phone?"' (Retrobrick, 2013).

The DynaTAC 8000X (Figure 6.3) weighed 800g, offered 30 minutes' talk time and 8 hours' standby and cost $4,000 (over $9,500, nearly £6,200, in 2014 money). It featured in the 1987 film *Wall Street* in which the main character, an American corporate raider named Gordon Gekko, was shown making a mobile phone call on the DynaTAC from a beach. Motorola's and other manufacturers' early portable cell phone designs were called 'brick' phones because of their size and weight, dictated by their use of 1G analogue radio communications.

A new model, the DynaTAC 8500X brick phone, arrived in the UK in 1987–8 costing £1,200 and quickly became a requirement for any 'yuppie' businessman.

In 1987 Nokia, which had been developing brick phones for the Nordic Mobile Telephone (NMT) service, released the Mobira Cityman. It cost the equivalent of over €4,500 and weighed 800g. It became known as the 'Gorba' when Soviet leader Mikhail Gorbachev was pictured in 1987 using a Cityman to make a call from Helsinki to his communications minister in Moscow (*Guardian*, 2010).

FIGURE 6.3 The first commercially available hand-held portable mobile phone, the Motorola DynaTAC 8000X, 1983. This is a 1984 model with a red LED display

CT2 phone point phones

An attempt to produce a smaller, lighter and less expensive 'mobile' phone – but which turned out to be a dead end – were the so-called CT2 phones launched in the UK in 1989. These were phones that could connect by radio to the ordinary telephone network when in the range of public 'telepoints' installed in some towns and cities. CT2 handsets could only be used to make outgoing calls, or used at home with a base station as a cordless phone to make and receive calls. *Which?* called the phones 'a pointless expense' (Consumers' Association, 1990) and the system was discontinued in the UK in 1993. However, CT2 survived longer in some other countries, for instance the French (two-way) CT2 service continued until 1997.

Digital phones

In the early 1990s there was a switch from 1G analogue to 2G digital cellular communications technology, which together with developments in electronics enabled much smaller and lighter handsets to be designed. The EU agreed a single digital standard for mobile phones across Europe – GSM – which gave European handset manufacturers, such as Nokia, a large home market, allowing them to expand rapidly. Nokia, a Finnish industrial conglomerate, which had taken a strategic decision in 1992 to establish a mobile communications division, launched the Nokia 1011, the first mass produced GSM 'candy bar' format phone (Figure 6.4).

The first mobile phone that combined voice calls with personal digital assistant (PDA) functions such as address book, calendar and the ability to receive emails and faxes, was the IBM Simon Personal Communicator introduced in 1993–4. The Simon also had a touchscreen interface operated using a stylus and is considered to be the first smartphone.

By the mid-1990s mobile phones were spreading to the mass market and had become desirable consumer products. Responding to this consumer demand, Motorola competed with its European rivals with the 1996 launch of the StarTAC, the world's first 'clamshell' phone (Figure 6.5) priced at $1,000. The StarTAC's design, which echoed the communicator device in the cult TV series *Star Trek*, was evolved from Motorola's 1989 MicroTAC analogue 'flip' phone with a hinged keyboard cover. Apart from the Sony CM-R111 clamshell design of 1994, which lacked a display but could fit into a large matchbox, the 88g StarTAC was the world's smallest and lightest phone and offered four hours of talk time and 47 hours' standby.

Nokia, which had decided by 1994 to focus on mobile communications for the consumer as well as business users, realised that mobile phones could be designed as fashion items as well as technical gadgets and, like Motorola, employed industrial designers as well as engineers (Kilpinen, 2013). In 1997 it released the Nokia 8110/8110i (Figure 6.6), which had a slide cover over the keypad and featured in the 1999 sci-fi film, *The Matrix*. When the film was released Nokia had already

FIGURE 6.4 The first mass-produced 'candy bar' mobile phone was the Nokia 1011 GSM digital phone, 1992. This is a 1993 model.

FIGURE 6.5 The first clamshell phone, the Motorola StarTAC, 1996

FIGURE 6.6 Desirable Nokia 8110i GSM slide phone, 1997, which featured in the film *The Matrix*

designed a fashionable new phone, the 7110, with a cover that slid open to answer a call at the touch of a button. It was also the first mass-market device that could use the new WAP mobile data service.

Nokia's 3210 and 3310 of 1999 and 2000 became iconic candy bar designs, which together sold nearly 200 million units (Figure 6.7a). Then in 2002 Nokia launched a mobile phone, the Nokia 1100, designed for users in developing countries and people only wanting a basic phone (Figure 6.7b). It had a dustproof cover, non-slip sides and a monochrome display and provided only the essential functions of making calls, sending text messages, alarm clock and calculator. It became the world's best-selling phone, with 250 million sold before it was discontinued in 1994.

Mobile phones   **129**

While the mobile phone companies were trying to introduce users to the early bulkier handsets that used 3G technology, Motorola's engineers and industrial design team created the very thin 2G RAZR, launched in 2004 (Figure 6.8). It quickly became the most desirable and fashionable phone to own, was available in several colours, including pink, and sold over 130 million worldwide.

Apple's original iPhone, launched in mid-2007, represented a breakthrough in handset design and technology. Two years earlier Apple's CEO, Steve Jobs, and his team had become excited with the idea of developing a phone that was much easier to use than existing designs. Two user interface ideas were pursued, one based on the iPod track-wheel, the second on a 'multi-touch' screen that was being developed for an Apple tablet computer (the forerunner to the iPad) together with a company called Fingerworks acquired by Apple. Jobs considered that a multi-touch interface, although technically riskier, had much more potential than the track-wheel, especially if built-in software functions like a touch-screen keyboard could be provided. A pre-launch prototype produced by the team under Jonathan Ive, Apple's head of design, was rejected by Jobs. He wanted a phone in which the specially developed tough 'gorilla glass' screen covered the whole front of the phone and dominated the form. Hence the phone's case and electronics had to be redesigned, delaying the iPhone's launch until June 2007 (Isaacson, 2011).

The original iPhone was designed to work on rather slow 2G networks, but Apple had already announced it was developing a 3G version and allowed third-party developers to start producing their own downloadable applications ('apps').

FIGURE 6.7a   Nokia 3310 GSM candy bar phone, 2000 was an iconic design that sold 126 million units worldwide

FIGURE 6.7b   Nokia 1100 basic GSM candy bar phone 2003, designed with a dustproof case for developing countries, sold about 250 million units

FIGURE 6.8   The highly fashionable Motorola RAZR V3 ultra-slim clamshell phone, 2004

In 2008 Apple launched the iPhone 3G (Figure 6.9 left) and introduced the App Store on its iTunes website for downloadable free and paid-for apps. Much of the functionality of the iPhone is provided by the apps on the phone and via the iTunes software on the user's personal computer, thus reducing the amount of electronics required in the handset itself. By the start of 2010 more than three billion apps had been downloaded by iPhone users across the world. Since Steve Jobs' death in 2011, Apple has continued to develop new models to attract new customers and encourage users to upgrade. The latest models compatible with the high-speed 4G networks being rolled out are the iPhone 6 and 6 Plus (Figure 6.8 right) launched in 2014, which had several improvements, including significantly larger 4.7-inch (11.9 cm) and 5.5-inch (14 cm) screens and thinner cases.

Other manufacturers soon brought out their own 3G/4G touchscreen smartphones to rival the iPhone, such as Samsung with its Galaxy range. These used different operating systems but also offered a large number of downloadable apps for every function a user might want or might not even have known they could have. Since the original iPhone, screens have steadily increased in size and several manufacturers have introduced 'phablets' – a smartphone and tablet computer hybrid with extra-large screens for watching video, playing games and basic office tasks.

FIGURE 6.9 The first touchscreen mobile phone was the 2G Apple iPhone, launched in 2007 with a 3.5-inch screen. (left) a 3G version of the original iPhone, 2008. (right) a 5.5-inch screen iPhone 6 Plus, 2014

Handset configurations

By 2013 there were essentially only three portable handset forms. The rectangular touchscreen phone had become the dominant design for smartphones (Figures 6.9a and 6.9b). A few smartphones with smaller screens for those who wanted a physical QWERTY keyboard were still made, notably by Blackberry (Figure 6.11), but were rapidly losing market share. Third, low-priced budget phones (some costing less than £10), usually of a simple candy bar form, such as the Nokia 100 (Figure 6.12), were marketed mainly for phoning and texting via 2G networks, but some also provided a camera, music player and/or FM radio (Which? Ltd., 2013b).

Andrew Muir Wood, who has researched the technical and design evolution of mobile phones, confirmed the divergence and convergence in mobile phone

(a) (b)

FIGURE 6.10 Dominant touchscreen smartphone designs
(a) 5.5-inch iPhone 6 Plus, 2014
(b) Samsung Galaxy S5, 2014

FIGURE 6.11 Blackberry Q10 smartphone with physical QWERTY keyboard, 2013

FIGURE 6.12 Nokia 100 budget phone, 2013

forms. From plotting the frequency with which different designs appeared in a technical magazine *T3* between 1996 and 2009, he found significant changes in the variety of designs on the market. In 1996 three-quarters of mobile phones were a simple candy bar form, then from 2002 clamshell, slide and touchscreen designs overtook the bar form and by 2009 about 60% of phones were touchscreen models (Figure 6.13, Muir Wood, 2010).

Since 2009, following the introduction of the iPhone in 2007, phone design has converged further with touchscreen phones becoming dominant, but Muir Wood found that the shift to simple rectangular forms had already occurred well before that. For example, a rectangular seamless metal body had featured before in successful products such as Sony Ericsson's T610 camera-phone in 2003, whose award-winning design was created by Swedish industrial designer, Erik Ahlgren (Figure 6.14).

Environmental impacts and regulation

When hand-held mobile phones were first introduced in the 1980s little was said about their environmental impacts. However, concern about the environmental impacts of mobile phones gradually emerged, initially focusing on the pollution and toxic hazards potentially arising at the phone's end-of-life, for example from the disposal to landfill of nickel cadmium batteries and tin-lead solder used in the production of the handset's electronics.

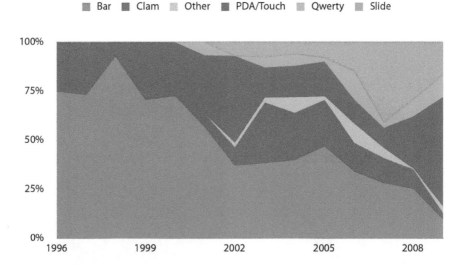

FIGURE 6.13 Mobile handset design forms appearing in a sample of the magazine *T3* (Tomorrow's Technology Today) between 1996 and 2009 showing the divergence and convergence of designs (From Muir Wood, 2010, p. 132)

FIGURE 6.14 Design award winning Sony Ericsson T610 camera-phone, 2003

End-of-life reuse and recycling

In 1997 a working group of major phone manufacturers, including Ericsson, Nokia, Panasonic and Philips, joined forces to respond to anticipated EU legislation on electronic product take-back, reuse and recycling. The group commissioned trials of take-back systems in the UK and Sweden to develop the industry's policy and practice. The trials found that only relatively few phones – almost none in Sweden – ended up in landfill. It also found that the phone recycling system was immature and that collecting used phones often produced more environmental impacts than recycling them (Gertsakis *et al.*, 1998).

In 2003 the Waste Electrical and Electronic Equipment (WEEE) Directive first came into force in EU Member States, followed by a revised version in 2014. The WEEE Directive set future collection, recycling and recovery targets for all types of electrical goods, including mobile phones. Within a few years many phone reuse and recycling schemes operated by manufacturers, network providers and charities, plus channels for selling used mobiles, had developed. But surveys showed that, although the replacement cycle for mobiles was getting shorter (under two years in Britain and the US and under three years in France, Germany and Japan), a high proportion of old mobiles were still kept as 'spares' or passed on to family or friends and only relatively few were recycled or sold. A 2012 study by Nokia showed that in industrialised countries 40% of old phones were kept, 18% given away, 12% recycled and 9% sold or traded in (Confused.com, 2012).

Hazardous substances

At the same time as the WEEE, the EU Restriction of Hazardous Substances (RoHS) Directive was enacted. This set restrictions for European manufacturers on the material content of new electronic equipment. European manufacturers, and exporters to Europe, responded to the Directive by changing some of the materials and chemicals used in phone production. RoHS required the phasing out of six substances used in the production of consumer electronic products put on the European market from July 2006 – four heavy metals (lead, mercury, cadmium, hexavalent chromium) and two categories of brominated flame retardants. Products that comply with RoHS are labelled with the 'CE' mark. Mobile phone batteries are covered by separate legislation, the EU Batteries Directive of 2006. This Directive prohibits batteries with a mercury or cadmium content above a certain threshold, and promotes the collection and recycling of waste batteries, in order to cut the amount of hazardous substances – in particular, mercury, cadmium and lead – dumped in the environment (EU, 2013).

Life-cycle impacts

While pollution and hazardous waste from mobile phone production and disposal is subject to environmental laws, at least in the EU, it gradually became recognised from the early 1990s that there were other major environmental impacts arising from the burgeoning production and sale of mobile phones. These impacts could only be uncovered by detailed life cycle assessment (LCA) studies. As was outlined in Chapter 2, LCA identifies different environmental impacts (greenhouse gas emissions, water pollution, solid waste, etc.) arising over the whole life cycle of a product, from raw materials extraction and processing through production, distribution and use, to end-of life reuse, recycling or disposal.

An LCA study for Nokia conducted in 1995 found that the main environmental impacts of a 3G mobile phone arose from the climate-changing carbon dioxide (CO_2) emissions generated by component manufacture during the production phase, and during the use phase from the energy used by the cellular radio network and the phone's charger on standby. According to Nokia's LCA, for a 3G phone over a 2-year life cycle, the emissions are relatively small; about 13.5 kg CO_2 per user for the handset alone (equivalent to the emissions of driving a medium car about 80 km) and 54 kg CO_2 per user if the energy use of the cellular network is included (equivalent to driving 315 km). Less significant impacts arose from the manufacture of circuits and displays; the international air transport of phones; and small, unmanaged releases of hazardous substances at end-of-life (Singhal, 1995).

Even before the European Commission's Integrated Product Policy Pilot Project (EC, 2005a; 2005b), which advocated LCA studies of mobile phones, Nokia began to publish detailed 'eco-profiles' of its handsets. For example, the eco-profile of the Nokia 100/1000 budget phone (Figure 6.12) published in 2011 listed the materials used in this 2G phone (mainly metals, ceramics and plastics, plus its battery and packaging) and which materials (e.g. PVC) were excluded to meet global regulations. The profile gave the phone's energy use when making a call, on standby and the charger's standby energy use when plugged in but not charging. The phone's 8 kg carbon dioxide equivalent (CO_2e) greenhouse gas emissions over a

3-year life-cycle, calculated from an LCA, showed that the phone's main impacts arise during product use (39% of its total CO_2e emissions), during raw materials acquisition and component manufacture (38%) and during transportation (17%). Relatively small impacts arose from Nokia factory operations (5%) and recycling (1%) (Nokia, 2011). These phone-related life-cycle emissions over three years, Nokia said, were equivalent to driving an average car just 53km, but not including emissions per user arising from operating the cellular network. A similar eco-profile for Nokia's Windows smartphone the Lumia 820 revealed 16 kg CO_2 equivalent emissions over three years, again excluding network operations (Nokia, 2012) – double the emissions of the budget Nokia 100. One of the unexpected results of these LCA studies was the significant amount of energy used during the use phase by phone chargers if left on standby. So, one of the measures that some manufacturers have taken to reduce emissions is to make more energy-efficient chargers, some of which give warnings to unplug when charging is finished.

Apple only began publishing environmental reports on its mobile phones in 2013. For example, the report on the iPhone 5S detailed its materials content and revealed that its life-cycle greenhouse gas emissions over three years were 75 kg CO_2e (this was more than reported for the Nokia smartphone, perhaps due to a more detailed LCA). Materials and manufacturing accounted for most (83%) of the iPhone's emissions. Partly of because the energy-efficiency of the charger, use of the phone produced only 12% of total emissions, with transport and recycling accounting for 5% and 1% respectively (Apple, 2014a).

Another way of expressing the life-cycle impacts of a product is its 'ecological rucksack'; meaning the weight of all the material resources (e.g. ores, rocks, fossil fuels, soil, water) extracted from nature at different phases of the product's life. This 'hidden weight' carried on the back of most products far exceeds the weight of the product itself. The ecological rucksack of a (non-smart) mobile phone has been calculated as 44 kg, more than 550 times more than the 80 g weight of the product itself. The mining and processing of aluminium, rare metals and other materials to make a phone uses the greatest amount of resources (27 kg); second the resources for electricity production in the use phase (including the phone network) account for 10 kg of the ecological rucksack; third the fuels, equipment, etc. for production of the phone involves 6 kg of resources; while the end-of-life phase disposal or recycling uses only 0.1 kg of resources (Welfens *et al.*, 2013).

Eco rating schemes

In 2010, O2 UK launched an eco rating scheme to inform its customers of the social and environmental impacts of different handsets as part of its 'Think Big' initiative (O2, 2010). The scheme was based not on a formal LCA, but on a 1 to 5 scoring system derived from answers by seven phone manufacturers (Nokia; HTC; LG; Samsung; Sony Ericsson; Blackberry; Palm) to a questionnaire, together with life-cycle information gathered by Forum for the Future, O2's partner in the project. The questionnaire covered:

1. Raw materials and manufacturing impacts: 26% of total eco rating score
2. Functionality (e.g. smartphone features): 25% of total score
3. Use impacts (excluding energy from device operation of the handset itself): 19% of total score
4. Corporate impacts (e.g. management systems; supply chain requirements; social inclusion and community; handset obsolescence): 11% of total score
5. Substance impacts: 7.5% of total score
6. Packaging and delivery: 7.5% of total score
7. Disposal impacts: 4% of total score.

Vodafone launched a similar 1 to 5 eco rating scheme for the handsets it supplies.

The eco rating scores for smartphones show that they have higher raw materials and manufacturing impacts than simple phones, but as multi-functional devices it is argued that they can combine many separate products in one and can facilitate more sustainable lifestyles, for example when used to plan routes for public transport, walking or cycling, or to allow remote online working.

In 2014 the eco rating systems used by O2 and Vodafone were replaced by a new scoring system, Eco Rating 2.0 (Forum for the Future, 2014). Eco Rating 2.0 was developed to combine the best elements of several rating systems. It is available to other network operators so that they share a common sustainability rating across all devices they offer.

Apple did not participate in the eco rating schemes. Instead, Apple's LCA studies showed that over 70% of the company's 33.8 million tonnes of greenhouse emissions in 2013 arose in the outsourced raw materials and production phases and 22% from consumer use of its products, with the rest coming from company offices, stores and data centres, transport and recycling. Apple therefore focused its efforts on the areas that it could control: auditing its suppliers' factories to ensure compliance with company codes on pollution and hazardous emissions and reducing greenhouse emissions by increasing the energy-efficiency of its products and using renewable energy to power company facilities (Apple, 2013; Apple, 2014b). In 2015 Apple announced it was to build two 20-megawatt solar farms in China and enable large areas of forest in the US to be protected, while supplying sustainable forestry products, for example for making packaging (Richard, 2015).

Ethical and political issues

In theory at least, Nokia's eco-profiles and the eco rating scores should help consumers to choose handsets according to certain ethical as well as environmental impacts. Other initiatives have focused mainly on the ethical issues, in particular improving conditions and eliminating child labour in phone factories and avoiding 'conflict minerals'. For example, in 2012 Apple said that it had acted to reduce excessive working hours in its main Chinese supplier and to avoid child labour and the use of conflict minerals (Apple, 2013). A BBC undercover investigation, however, criticised Apple for failing to meet its commitments on working hours and labour conditions in at least one of its Chinese suppliers and for sourcing tin from illegal mines in Indonesia (Bilton, 2014).

Conflict minerals refers to materials used in phone manufacture, such as tin, tantalum, tungsten and gold, which are mined and traded under the control of armed groups operating in certain African states (notably the Democratic Republic of Congo) who violate human rights through forced labour, rape and killings. This issue has been publicised notably in journalism with titles like 'My Search For A Smartphone That Is Not Soaked In Blood' (Monbiot, 2013), but there is also internationally agreed guidance for companies to help them avoid supply chains that trade in conflict minerals (OECD, 2013). However, because the origin of conflict minerals is hard to identify and they are often mixed with legitimately mined minerals, the existing eco rating schemes recognise, but do not include, conflict minerals in their assessments.

In addition to suppliers' rating schemes and critical journalism, the impacts of the production and use of mobile phones have had increasing attention from campaigning organisations. For example, the Gaia Foundation published a highly critical report on the effects of extracting the raw materials required to make electronic consumer goods, including mobile phones. Gaia criticised the mining and other operations involved for depleting natural resources; pollution; toxic waste and land grabs; use of conflict minerals and poor working conditions. It also criticised the resulting products for their in-built obsolescence (Sibaud with the Gaia Foundation, 2013). Perhaps ironically, Gaia's report was publicised via a spoof mobile phone advertisement disseminated via Twitter (Figure 6.15). The Gaia report also advocated solutions to the problems it identified, such as the development of a 'circular economy', a concept originally proposed by the Ellen MacArthur Foundation (2013), in which products are reused, remanufactured and recycled in order to achieve zero waste. Other organisations are attempting more immediate solutions, such as the development by a Dutch social enterprise of an ethical Android smartphone called 'Fairphone', originally conceived to avoid use of conflict minerals. However, by 2014, because of the labyrinth of suppliers involved, only two of the phone's many minerals, tin and tantalum, came from validated conflict-free mines. Among its other requirements, the Fairphone was designed to be repairable to extend its useful life, recyclable and manufactured in Chinese factories with fair pay and good working conditions. It was funded via pre-orders from the public and launched in 2013, with about 50,000 phones sold by 2014 (Fairphone, 2014).

Social influences and impacts

Few people could have predicted the enormous effect that mobile phones would have on society. There is not space to discuss these effects in detail, but they range from facilitating political revolutions in Arab states to the explosion in people's use of Facebook, Twitter and other social networking systems. Mobile phones have become an everyday companion for most people. Users get the feeling of the world in their hand and being connected to other people wherever they

FIGURE 6.15 Spoof advertisement highlighting the negative impacts involved in the extraction of materials for making mobile phones. The 'apps' on this phone are: resource depletion, ecosystem destruction, land grabbing, inbuilt obsolescence, toxic waste, conflict minerals, poor working conditions

are, meaning that the mobile phone has become an integral part of people's lives (Welfens *et al*., 2013). A 2013 Ofcom report found that in 2012 over half of UK adults used a mobile phone to go online, rising to 86% among smartphone users. Nearly three-quarters of those with a social network profile claimed to visit their websites at least daily; half claimed to visit the sites more than once a day, with nearly 10% visiting more than ten times a day (Ofcom, 2013b). A previous Ofcom report observed that the UK was a nation becoming 'addicted to smartphones'. It found that over a third of adult smartphone users admitted high levels of addiction to their phone, rising to 60% of teenage users. The report commented that smartphones are changing social habits and etiquette. Over half of adults and two-thirds of teenagers said they had used their smartphone while socialising with others, nearly a quarter of adults and a third of teenagers had used their phone during mealtimes and over a fifth of adults and nearly half of teenage smartphone users admitted using or answering their handset in the bathroom or toilet. The growing functionality of smartphones, Ofcom found, is also affecting people's other leisure activities. Over half of UK adult smartphone users claimed to be doing less of other activities, while over two-thirds of teenage users said they did some activities less than before, such as watching TV (23%), and reading books (15%) (Ofcom, 2011b). A more recent report found that, on average, British 16- to 24-year-olds spent over three and a half hours per day on a mobile phone, of which one and a half hours per day was accessing social media (Ofcom, 2014). The Ofcom surveys also showed that smartphones are also merging people's home and work lives. Among users who work, 30% regularly made personal calls at work on their smartphone, while 35% regularly use their phone for work calls while 'off duty', and a quarter said they use their phone for work while on leave (Ofcom, 2011b). International surveys show that a quarter to a third of mobile subscribers in many countries have two or more handsets – Germany (25%), the UK (29%) and China (36%) – and in these countries over 70% of voice calls were made by mobiles (Ofcom, 2011b).

These behavioural changes have inevitably led to critical comment, such as that of the eminent neuroscientist, Baroness Susan Greenfield, who is researching the effects of increasing exposure to screen technologies on the human brain and said:

> Already, it's pretty clear that the screen-based, two dimensional world that so many teenagers – and a growing number of adults – choose to inhabit is producing changes in behaviour. Attention spans are shorter, personal communication skills are reduced and there's a marked reduction in the ability to think abstractly.
>
> *(Greenfield, 2015)*

The author and journalist, Tim Lott (2015), wrote in his family column, 'Our intoxication with new technology is eroding bonds, including family bonds, even as it re-makes them in a different form ... I walked into the kitchen to see three family members each with not one but two screens operating in front of them (tablets and phones) as they exchanged occasional conversational gambits.' Others argue, however, that despite the dangers of 'sexting', invasion of privacy and online bullying, for most teenagers heavy use of mobile phones for texting and social networking is not generally harmful. On the contrary, mobile phones and computers can facilitate social interaction, both online and face-to-face, teach people to use information and give them the opportunity to write and publish their ideas to their friends or to the world (Thompson, 2013). For example, a study of cell-phone use among American teenagers found that they mainly accessed their phones for remote chatting, meeting up with friends and sharing images (boyd, 2014).

Consumer behaviour and preferences have of course also influenced mobile phone systems and handset design. One of the most striking examples was the unexpected growth in demand for text messaging following the introduction of 2G digital mobile systems in the 1990s. Initial growth was slow, with global GSM users in 1995 sending on average only 0.4 messages per month. By the end of

2000, the average number of messages had risen to 35 per user per month. By 2010, 6.1 trillion SMS text messages were sent (Wikipedia, 2013b), while in the UK alone 129 billion text messages were sent, an average of five text messages a day for every person (Ofcom, 2011b).

Changes in handset interface design influenced the growth of text messaging. From small keypads in which each key accessed several letters and numbers, thus requiring multiple key presses to compose a message and the emergence of a transient shorthand SMS 'language' (e.g. CU@8 for 'see you at 8 o'clock'), the mobile phone evolved to provide full touchscreen keyboards. One reason for the popularity of Blackberry phones among business users and teenagers before the widespread use of smartphones was their full physical keyboard and their secure proprietary Blackberry Messenger system for free texting via the Internet. The Messenger system, for example, was reported to have been used extensively by young people to communicate during the August 2011 riots in London.

As noted earlier, the increasing use of smartphones for applications other than phone calls and texts, such as watching video or TV, playing games, drawing and painting, online learning, etc. has led to larger screen sizes. For example, the iPhone 6 Plus has a 5.5-inch (14 cm) diagonal screen compared to 3.5-inch (8.9 cm) for iPhones 3 and 4 (see Figure 6.8).

Future developments

Because of the rapid changes in technology, the future of mobile phones is hard to predict. Some developments that are being introduced or anticipated include use of phones for instant cash payments, phones with flexible screens, and phones recharged by movements of the user or by solar cells (Institute of Physics, 2013). Existing smartphones include sensors such as accelerometers and compasses. Future smartphones are likely to have even more sensors built into them; for example, they could detect ambient temperature and other environmental conditions, to provide a personalised weather forecast for a local area. Some models already include software that enables the phone's camera to sense, look up or provide information about objects or buildings it is pointed at. Mobile phones with relevant apps have already become the remote interface and controller for other products, including 'smart' lights, washing machines and television sets and this is likely to continue.

In mobile phone design there have been radical concepts such as the Nokia Morph with a flexible screen (Nokia Research, 2013) and there is evidence of the beginnings of a divergence in design away from the dominant rectangular touchscreen form. One new design concept is 'Phonebloks', a modular phone made up from component blocks such as processor, screen and battery, which can be upgraded or replaced if broken, thus extending the life of the product (Phoneblok, 2015). Another example is the development of smartwatches. These are tiny computers, most of which are accessories wirelessly linked to a compatible smartphone to provide functions such as alerting the user to reminders, sending and receiving texts, making and receiving calls, taking photos, sports monitoring and running a limited number of apps, all from the user's wrist. A *Which?* review of the new Samsung Galaxy Gear smartwatch concluded that, at nearly £300, it was probably not worth buying because of the inherent problems of the watch format. For example, you had to hold the watch to your face for phone calls and the screen was too small to display more than a few sentences without scrolling (Which? Ltd., 2013c). But the review said it was too early to give a verdict on this fast-evolving 'wearable' technology and a later review acknowledged that smartwatches could help avoid missing messages and calls and be more convenient for some uses than a smartphone (Which? Ltd., 2014).

The next generation of mobile networks, 5G, is due to be introduced around 2020, offering communication speeds up to ten times those provided by 4G, so for example allowing a feature film to be downloaded in a few seconds or to enable users to play online games with high graphics content in high-definition.

Because mobile communications technology is evolving so fast, highly speculative ideas are being explored by some scholars of the future. An example is how today's wearable products – such as the prototype Google Glass headset (Figure 6.16) which is connected to a smartphone to make calls, take photographs and search Google with images displayed on a screen in front of one eye – might evolve. Such products could perhaps be developed into an implanted electronic communication device, such as that portrayed in the 2012 dystopian science-fiction film *Total Recall*, in which the user's palm becomes the keyboard and the device is powered by the person's movement (Fawcett, 2014).

The pattern of innovation

The mobile phone is the outcome of nearly 70 years of continuous technological innovation and design evolution beginning just after the Second World War. It

FIGURE 6.16 Google Glass with an optical head-mounted display, providing many computer and communications functions in a wearable product, especially when paired with a smartphone. It was available from 2013 to 2015 when the 'Project Glass' experimental phase ended and new versions were under development

has involved not only product innovations, but also innovations in services and systems. The product innovations include the miniaturisation of electronic components; smaller and longer-life batteries; and major improvements in user interface design and in operating system and application software, all of which have been embodied in many different designs of mobile handset. The service and systems innovations range from text messaging and different tariffs for owning and using phones to the successive generations of radio communications technology.

Once cellular communications technology had been established in the late 1970s it took over 25 years for the touchscreen smartphone using 3G or 4G cellular communications to become the dominant design. Starting from the analogue 'brick', mobile phone design diverged into several handset configurations (candy-bar, clamshell, slide, etc.), made possible by digital communications and increasingly powerful, efficient and compact electronics and batteries, before converging onto the touchscreen smartphone. Despite this currently dominant design, innovation is continuing, for example towards 5G communications and 'wearable' products such as smart watches.

References

Apple (2013) Apple supplier responsibility. 2013 Progress Report, Apple Inc., January.
Apple (2014a) iPhone 5s Environmental Report, Apple Inc., September.
Apple (2014b) Apple environmental responsibility, 2014 progress report, Apple Inc.,
Baird, D. (2013) Mobile phones: 40 years of handsets in pictures [Online]. Available at www.guardian.co.uk, 3 April (accessed August 2013).
Bilton, R. (2014) Apple failing to protect Chinese factory workers [Online], BBC Panorama, 18 December. Available at http://www.bbc.co.uk/news/business-30532463 (accessed February 2015).
boyd, d. (2014) It's Complicated: The Social Lives of Networked Teens [Online]. Available at http://www.danah.org/itscomplicated/ (accessed September 2014).
Confused.com (2012) What happens to old mobile phones? [Online]. Available at http://www.confused.com/news-views/infographics/old-mobile-phones-and-e-waste (accessed February 2014).
Consumers' Association (1990) 'A pointless expense?' *Which?* February, p. 75.
Consumers' Association (1994) 'How to buy a mobile phone', *Which?* July, pp. 44–49.
EC (2005a) Stage 1 Final Report: Life Cycle Environmental Issues of Mobile Phones [Online]. Available at http://ec.europa.eu/environment/ipp/mobile.htm (accessed August 2013).
EC (2005b) Stage 2 Final Report: Options for Improving Life-Cycle Environmental Performance of Mobile Phones [Online]. Available at http://ec.europa.eu/environment/ipp/pdf/nokia_st_II_final_report.pdf (accessed August 2013).
Ellen MacArthur Foundation (2013) In-depth: mobile phones [Online]. Available at http://www.ellenmacarthurfoundation.org/business/toolkit/in-depth-mobile-phones (accessed September 2013).
EU (2013) Summaries of European legislation – waste management [Online]. Available at http://europa.eu/legislation_summaries/environment/waste_management/index_en.htm (accessed August 2013).

Fairphone (2014) Fairphone, [Online]. Available at http://www.fairphone.com/ (accessed September 2014).

Fawcett, K. (2014) The Future is Here: What's Next For Mobile Phones? [Online]. Available at http://www.smithsonianmag.com/smithsonian-institution/the-future-is-here-whats-next-for-mobile-phones-180951479/?no-ist (accessed September 2014).

Forum for the Future (2014) Eco Rating 2.0. Introduction to the Tool, September [Online]. Available at http://www.forumforthefuture.org/sites/default/files/files/Introduction%20to%20Eco%20Rating%202_0(1).pdf (accessed February 2015).

Guardian (2010) From bricks to the iPhone: 25 years of the mobile phones, guardian.co.uk, 14 February [Online]. Available at http://www.theguardian.com/technology/gallery/2010/feb/14/mobile-phones-gadgets-iphone (accessed August 2013).

Guardian (2013) Mobile phone's 40th anniversary: from 'bricks' to clicks, guardian.co.uk, 3 April [Online]. Available at http://www.guardian.co.uk/technology/2013/apr/03/mobile-phone-40th-anniversary (accessed September 2013).

Gertsakis, J., Morelli, N. and Ryan, C. (1998) *An introduction to extended producer responsibility*, Melbourne, Australia. Centre for Design at RMIT, August.

Greenfield, S. (2015) Screen technologies [Online]. Available at http://www.susangreenfield.com/science/screen-technologies/ (accessed February 2015).

IDC (2014) Worldwide Quarterly Mobile Phone Tracker, International Data Corporation, Framingham, MA, November [Online]. Available at http://www.idc.com/getdoc.jsp?containerId=prUS24461213 (accessed September 2014).

Institute of Physics (2013) Future mobile phone technology [Online]. Available at http://www.physics.org/article-questions.asp?id=83 (accessed September 2013).

Isaacson, W. (2011) *Steve Jobs*, London, Little Brown.

Kilpinen, P. (2013) *Capability Development within the Multinational Corporation*, Aalto University publication series, Doctoral dissertations 13/2013, Helsinki, Finland, Aalto University School of Business.

Lott, T. (2015) 'An electronic apocalypse is coming unless we act now', *The Guardian*, 9 January.

Monbiot, G. (2013) 'My Search For A Smartphone That Is Not Soaked In Blood', Z-Net 13 March [Online]. Available at http://www.theguardian.com/commentisfree/2013/mar/11/search-smartphone-soaked-blood (accessed September 2013).

Muir Wood, A. (2010) 'The Nature of Change in Product Design. Integrating Aesthetic and Technical Perspectives', PhD thesis, Design Management Group, Institute for Manufacturing, Cambridge, University of Cambridge, September.

Nokia (2011) Nokia 100/1000 Eco-profile [Online]. Available at http://www.nokia.com/LCA (accessed August 2013).

Nokia (2012) Nokia Lumia 820 Eco-profile [Online]. Available at http://www.nokia.com/LCA (accessed August 2013).

Nokia Research (2013) The Morph concept [Online]. Available at http://research.nokia.com/morph (accessed September 2013).

O2 (2010) O2 eco rating, London, Forum for the Future and O2 [Online]. Available at http://www.o2.co.uk/assets2/thinkbig/O2Ecoratingbrief_Aug2010.pdf (accessed January 2015).

OECD (2013) *Due diligence guidance for responsible supply chains of minerals from conflict-affected and high-risk areas* (2nd edn), Paris, Organisation for Economic Co-operation and Development.

Ofcom (2011a) ICMR 2011 telecoms key market developments charts, Ofcom, London, November [Online]. Available at http://stakeholders.ofcom.org.uk/binaries/research/cmr/cmr11/icmr/ICMR2011.pdf (accessed September 2013).

Ofcom (2011b) The communications market 2011, London, Ofcom [Online]. Available at http://stakeholders.ofcom.org.uk/market-data-research/market-data/communications-market-reports/cmr11/ (accessed August 2013).

Ofcom (2013a) International Communications Market Report, London, Ofcom, December [Online]. Available at http://stakeholders.ofcom.org.uk/binaries/research/cmr/cmr13/icmr/ICMR_2013_final.pdf (accessed September 2014).

Ofcom (2013b) Adults' media use and attitudes, London, Ofcom, April [Online]. Available at http://stakeholders.ofcom.org.uk/market-data-research/media-literacy/archive/medlitpub/medlitpubrss/adults-media-use-attitudes/ (accessed August 2013).

Ofcom (2013c) 4G, London, Ofcom [Online]. Available at http://consumers.ofcom.org.uk/4g/ July (accessed August 2013).

Ofcom (2014) Communications market report 2014, London, August [Online]. Available at http://stakeholders.ofcom.org.uk/binaries/research/cmr/cmr14/2014_UK_CMR.pdf (accessed September 2014).

Phonebloks (2015) Phonebloks: a phone worth keeping [Online]. Available at https://phoneblocks.com/en (accessed July 2015).

Retrobrick (2013) Retro Brick: Motorola DynaTAC 8000x, UK [Online]. Available at http://www.retrobrick.com/moto8000.html (accessed August 2013).

Richard, M.G. (2015) Apple buys a forest, will build 2 new solar farms [Online]. Available at http://www.treehugger.com/corporate-responsibility/apple-buys-forest-size-sanfrancisco-36000-acres-conservation-will-build-2-new-solar-farms-china.html (accessed May 2015).

Sibaud, P. with Gaia Foundation (2013) Short circuit: The lifecycle of our electronic gadgets and the true cost to the Earth, London, Gaia Foundation [Online]. Available at http://www.gaiafoundation.org/short-circuit-the-lifecycle-of-our-gadgets-and-the-true-cost-to-earth (accessed August 2013).

Singhal, P. (1995) Life cycle environmental issues of mobile phones, Integrated Product Policy Stage 1 Final Report, Espoo, Finland, Nokia Corporation, April [Online]. Available at http://ec.europa.eu/environment/ipp/pdf/nokia_mobile_05_04.pdf (accessed August 2013).

Skipworth, H. (2013) The most iconic mobile phones in history, celebrating 40 years since the first call, *Pocket-lint* 3 April [Online]. Available at http://www.pocket-lint.com/news/120303-mobile-phone-40-year-anniversary (accessed October 2013).

Thompson, C. (2013) *Smarter than you think: how technology is changing our minds for the better*, London, William Collins.

Web Design (2009) The evolution of cell phone design between 1983–2009, *Web Design*, 22 May 2009 [Online]. Available at http://www.webdesignerdepot.com/2009/05/the-evolution-of-cell-phone-design-between-1983-2009/ (accessed October 2013).

Welfens, M.J., Nordmann, J., Seibt, A. and Schmitt, M. (2013) Acceptance of mobile phone return programmes for increased resource efficiency by young people, *Resources*, vol. 2, pp. 385–405.

Which? Ltd. (2005a) 'Mobile phones', *Which*? May, pp. 38–43.

Which? Ltd. (2005b) 'Mobile phones', *Which*? December, pp. 56–60.

Which? Ltd. (2013a) 'Testlab Smartphones', *Which?* August, pp. 42–44.

Which? Ltd. (2013b) Mobile phone reviews [Online]. Available at http://www.which.co.uk/technology/phones/reviews/mobile-phones/ (accessed August 2013).

Which? Ltd., (2013c) 'Samsung launches smartwatch', *Which?* October, p. 12.

Which? Ltd., (2014) 'What's the point of a smartwatch?', *Which?* April, pp. 44–45.

Wikipedia (2013a) History of mobile phones [Online]. Available at http://en.wikipedia.org/wiki/History_of_mobile_phones (accessed August 2013).

Wikipedia (2013b) Text messaging [Online]. Available at http://en.wikipedia.org/wiki/Text_messaging (accessed August 2013).

University of Salford (2013) Computer Networking and Telecommunications Research, School of Computing, Science & Engineering [Online]. Available at http://www.cntr.salford.ac.uk/comms/etacs_mobiles.php (accessed February 2013).

7
LESSONS FOR PRODUCT DESIGNERS, DEVELOPERS AND INNOVATORS

In this chapter I will draw out conclusions from the case studies in the previous chapters and attempt to provide some practical lessons for all those involved in the planning, development and introduction of new products and innovations. The chapter will also try to relate the empirical evidence of the case studies to existing theories of product innovation, design evolution and environmental sustainability.

Conclusions and lessons similar to several of those in this chapter may be found in the literature on business strategy, new product development, innovation management, marketing and sustainable design; for example, in standard works by Porter (1990), Brezet (1998), Rogers (1995), Kotler (2000), Cooper (2011) and Tidd and Bessant (2013). This chapter provides further evidence to support many of these well-known conclusions and lessons, but also a number of others derived from the empirical evidence of the product case studies.

Case study examples to illustrate the book's general conclusions are shown in boxes within sections and the practical lessons are shown in **bold** type, at the ends of sections and sub-sections. The conclusions and lessons are intended to be of relevance to designers, engineers, product planners and educators in these fields, as well as to inventors, business strategists, marketers and interested consumers.

Understand patterns of innovation

The first conclusion is that the different consumer products follow patterns of innovation, design and evolution, going through one or more divergent, convergent and divergent phases, as exemplified by the pattern of bicycle evolution in Chapter 1. However, there are differences depending on the technologies on which the products are based, as will be shown in this chapter.

148 Lessons for product designers, developers and innovators

Early divergent experimentation

For all the products, one or more key inventions were created that started an initial divergent phase of design experimentation and technical development. These diverse early designs might be based on the same or different technologies and may use long-established or new components and materials. Early designs often look like an assembly of functional parts, but which become increasingly integrated as the parts are enclosed and the product is designed as a whole. This phase of evolution is typically driven by the attempts by inventors, designers, engineers and manufacturers to eliminate the deficiencies of existing designs and produce more practical and desirable products. Socio-economic, commercial, and sometimes political, forces also affect the evolutionary process. In an influential publication, Utterback and Abernathy (1975) described this as the Stage I or 'fluid' phase of innovation (Figure 7.1).

Throughout this chapter the general conclusions and lessons will be illustrated with examples taken from the product case studies and shown in boxes. Below is the first set of examples to illustrate the early divergent phase of innovation.

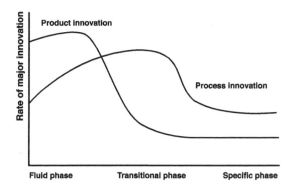

FIGURE 7.1 Typical pattern of product and process innovation over time. (Adapted from Utterback and Abernathy (1975), p. 645 and Abernathy (1978), p. 72.)

Washing machines

The first electric washing machines had wooden tubs and mechanisms similar to hand-operated machines, but used hazardous externally-mounted electric motors to agitate the wash (Chapter 2, Figures 2.3 and 2.4). Electric washing machines then evolved from these early twentieth-century designs via a wide variety of forms and agitating mechanisms (paddles, tumbling, tilting, vibrating, etc.) towards today's top- and front-loading designs introduced just before and after the Second World War. The drivers of innovation were initially to electrify a manual device and then to improve its cleaning and drying performance, safety, appearance and convenience.

Television

Low-definition, electro-mechanical television – notably the system invented by Baird in the 1920s – was demonstrated and used for early broadcasts (Chapter 4, Figure 4.1). By the 1930s electro-mechanical systems began to be displaced by higher-definition, less cumbersome all-electronic systems using cathode ray tubes, invented and introduced by Farnsworth Television and RCA in the US and EMI in Britain. Post-war black and white, and subsequently colour television, then continued to be improved with higher-definition pictures, larger screens and less bulky cabinets or casings. The drivers of innovation were: the search for higher-definition image capture, transmission and reception; the need for a less complex and cumbersome television system; and affordable domestic TV sets that could display colour as well as black and white pictures.

Mobile phones

Car radio telephones with a limited range were introduced in the mid-1940s. Despite improvements to communications technology, these fixed car and bulky 'transportable' radio phones could be used by only a few subscribers at a time. The evolution towards today's mobile phone systems began when the US government allowed use of the radio frequencies required for practical application of the 1947 theoretical concept of cellular communications, which would enable many users to make calls simultaneously. This led in the 1970s to the development of the first experimental cellular radio networks and the first portable cellular phone, launched by Motorola in 1983 (Chapter 6, Figure 6.3). The drivers of change were: to develop and apply innovations in communications technology and electronics to overcome the user limitations of early radio phones and to increase their portability, performance, desirability and usefulness.

The emergence of dominant designs

Following this early divergent experimental phase, one or more dominant designs typically emerge – as originally suggested by Utterback and Abernathy (1975), with the term specifically coined by Abernathy (1978). The dominant design(s) first appear in the Stage II or 'transitional' phase and become established in the Stage III or 'specific' phase in their innovation models (Figure 7.1).

The dominant product's design converges on one or more technologies and configurations, for example the standard diamond-frame bicycle, GLS incandescent light bulb or conventional upright and cylinder vacuum cleaners. Instead of the divergent invention and technical experimentation of the fluid phase, the efforts of designers, engineers and manufacturers focus on making incremental product improvements and stylistic changes and on introducing new or improved components, materials and production processes. The changes in this phase are driven by continued attempts to improve and eliminate the shortcomings of existing designs, to reduce production costs and respond to customer feedback and changing fashions. Manufacturers often create products for different market segments and may start designing to reduce environmental impacts. This results in ranges of products – often variants on a common product platform – offering specifications, performance and prices that can dominate the market. Once established, the dominant product configuration gains momentum and follows a set of constrained possibilities within what Dosi (1982) has called a 'technological trajectory'. It becomes difficult to introduce major innovations outside of this trajectory, because most technical knowhow and development resources are focused within it.

Washing machines

Just before the Second World War the two configurations that would become dominant – top-loading vertical axis and automatic front-loading horizontal axis machines – were introduced. The designs also converged from a variety of cylindrical and sink-shaped machines with enamelled steel cabinets, often on legs, into today's familiar white boxes. However, it took until the 1980s before fully automatic designs had largely displaced manual or semi-automatic single- and twin-tub top-loading types. The basic form of automatic washing machines then remained unchanged, but with many improvements and in a wide variety of models for different markets (Chapter 2, Figures 2.8 and 2.9).

Television

The TV set with a cathode ray tube (CRT) screen mounted horizontally in a simple wooden or plastic cabinet became the dominant design after the Second World War (Chapter 4, Figures 4.7b and 4.7c). This configuration displaced any surviving electro-mechanical and mirror-lid electronic receivers. This basic configuration survived with screens of increasing size until the arrival of flat panel TVs in the early twenty-first century. However, there were many significant technical innovations, including colour replacing black and white CRTs, solid state electronics replacing valves, and digital replacing analogue tuners. In addition, incremental improvements such as remote controls, and designing for economic manufacture and easy servicing were introduced. At the same time CRT TV sets could be linked to an increasing number of accessories, such as videocassette recorders (VCRs), DVD players and games boxes, and new services such as satellite TV and teletext.

> **Mobile phones**
>
> The design of mobile phone handsets depended to a great extent on the rapid innovation in electronics and communication systems. As cellular communications evolved from analogue to successive generations of digital systems, and integrated circuits became more powerful and miniaturised, mobile phones evolved in design from heavy 'brick' models into a wide variety of smaller and lighter candy-bar, clamshell and slide designs, including phones in fashionable colours and with computer-like functions. Almost all mobile phones began to be built around the tiny power- and component-saving microprocessor chips designed since 1993 by the British company ARM. By 2013 rectangular phones with a large touch-screen, pioneered by the first iPhone, had become the dominant design for smartphones (Chapter 6 Figure 6.9). Smartphones with a physical keyboard, notably those designed by Blackberry, were still produced but were losing market share, while small-screen budget mobiles were sold mainly for phoning and texting.

Innovative design

The dominant design phase is typically followed by another period of technological divergence and design variety. This phase arises because inventors, designers, engineers, manufacturers and new entrants to the market start to apply new product or process technologies, materials and components to create radically new products and innovations. These new technologies, materials and components may also be applied to reduce costs and improve the performance, reliability, aesthetics and emotional appeal of existing products. An important driver for the development of these innovative and improved products is to cope with stagnating or saturated consumer demand. Innovation is also required to fend off competition from low-cost manufacturers, to generate new consumer wants, and to meet environmental and safely legislation or standards.

This second divergent phase does not feature in Utterback and Abernathy's (1975) or Abernathy's (1978) innovation models (Figure 7.1). However, it can be partly explained by innovation S-Curve theory (Foster, 1988) and the concepts of *sustaining* versus *disruptive* technologies (Utterback, 1994; Christensen, 2000).

The S-Curve represents product performance as a function of time or effort. The curve indicates that after an initial slow period of development, a dominant design is established and there are rapid improvements using performance-enhancing 'sustaining' technologies, which eventually level off as further improvement becomes increasingly difficult. At this point one or more products based on a completely new 'disruptive' technology – products which up to then may have had inferior performance – can equal and eventually overtake the performance of the dominant design. This overtaking may occur even if there is a burst of improvements in the dominant design in an attempt to fight off competition from the invading disruptive products (Figure 7.2). Innovation theorists such as Foster and Christensen suggest that disruptive innovations usually come from creative small businesses like Dyson's company (Chapter 5) trying to enter an industry dominated by large corporations. But the evidence of this book's case studies is that disruptive innovations are often developed by multinationals with large R&D resources, although they may use inventions and innovations created by small businesses, as happened with Apple's iPhone (Chapter 6).

Whatever the source, the technological competition seen in the early experimental phase of product innovation reappears in the second divergent phase. In this phase dominant designs may survive for a long time alongside the innovative products, as incandescent light bulbs did in competition with fluorescent lamps, or may disappear, as 1G analogue mobile phones did when 2G and 3G digital phones were introduced.

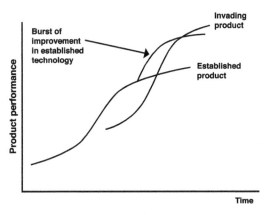

FIGURE 7.2 Technological S-Curve: diminishing improvements in performance over time result in competition between, and eventual displacement of, existing dominant designs by innovative products based on a disruptive technology. (Adapted from Utterback (1994), p. 160)

Washing machines

Automatic washing machines remained essentially unchanged from the 1960s, but with improvements that focused initially on increasing spin speed, reducing production costs and improving reliability. Further improvements included smart electronic controls, wash programs for low-temperature detergents and machines with reduced energy, water and detergent use. However, active R&D on washing machines continued and several innovative designs were introduced in the early twenty-first century. These included Dyson's Contrarotator (Chapter 2, Figure 2.13b), water electrolysing and bead washing machines (Chapter 2, Figure 2.14) that offered or offer further reductions in energy, water and detergent use. But so far these innovative machines have not displaced the dominant designs. The main drivers behind these improvements and innovations were: competition for improved wash performance plus environmental standards and legislation requiring reduced energy and water use.

Lamps and lighting

From the late 1950s lighting technology has followed a steady divergent path from the dominant GLS incandescent light bulb to include more efficient fluorescent and halogen incandescent lamps. It then diverged further to include even more efficient compact fluorescent (CFL) and solid state (LED) lamps in various shapes and sizes, before converging slightly with the withdrawal of GLS incandescent light bulbs from many countries. Because of their improving efficiency, tungsten halogen incandescent lamps will continue to compete for some time with CFL and LED technologies. The drivers behind the many innovations in lighting had similar technical and commercial goals – increased technical efficiency and durability at acceptable cost, stimulated by the need to reduce environmental impacts and gain commercial advantage.

Television

From the early twenty-first century television equipment design diverged to include LCD, plasma and LED flat panel screens, digital tuners, 3D pictures and high-definition displays driven by powerful microprocessors. These new products have almost fully displaced the formerly dominant CRT television. At the same time DVD and hard disk recorders and Internet streaming have displaced videocassette recorders. The continued technical evolution of television equipment is nicely illustrated by a *Which?* report that compared Best Buy TV sets of 2006, 2009 and 2014. The report observed that during that time TVs had gone through many changes, getting slimmer and with larger, almost frameless screens (Chapter 4, Figure 4.9b). Screen resolution had improved to full HD and ultra-HD; LED TVs had replaced both earlier LCD and plasma screens; sound quality had improved and smart TVs with Internet access had become commonplace. At the same time energy use had halved for comparable size sets (Which? Ltd., 2014a). The driver of these improvements and innovations is the intense technological competition between the major manufacturers aimed at encouraging consumers to replace their equipment with higher specification, more profitable models and to respond to standards and legislation on energy and materials use.

Sources of creative ideas

The creative ideas behind the inventions and new designs produced by inventors, engineers and designers at the different phases of product evolution are the result of various triggers and sources. The creative trigger is often a personal need, unmet market demand, or deficiency in existing products perceived by an inventor, manufacturer or marketer. The creative idea itself may come from taking an existing technology or product in one field and transferring or adapting it to another; it may be from thinking of an analogy between one problem and another; or it may be the result of combining ideas, designs or technologies from different fields to create something new (see for example Roy (2013)).

Vacuum cleaners

The Bissells' idea for the carpet sweeper arose from their need to sweep bits of packing straw from the carpets of their china shop. Spangler's invention of the electric vacuum cleaner (Chapter 5, Figure 5.3a) arose from the dust created by using a Bissell carpet sweeper aggravating his asthma and he had the idea after observing the action of a rotary street sweeping machine. Hoover's engineers wanted to improve the performance of Spangler's cleaner and adapted the idea of a carpet beater by adding beaters to the rotating front brush. Dyson's dissatisfaction with the declining suction of existing bagged vacuum cleaners led him to invent and design a bagless cyclonic cleaner, the idea for which arose from transferring industrial cyclone technology to a domestic cleaner (Chapter 5 Figures 5.10a and 5.11a). Robotic vacuum cleaners have been made possible by transferring and adapting industrial and recreational robotic and other technologies to domestic cleaners (Chapter 5, Figure 5.14).

Different patterns of innovation

Although the above sections reveal patterns of divergence, convergence and divergence for all the case study products, there are clear differences in the rates and extent of change depending on the technologies on which the products are based – as outlined in the 'pattern of innovation' summaries at the end of the previous chapters. Washing machines, an electro-mechanical product, are still in an early second divergent phase with the established designs still dominant. Electric lighting has been in a divergent phase for decades with new technologies challenging the dominant design and is about to converge again towards a new dominant design. The electronic products, television and mobile phones, have already passed through at least two divergent and convergent phases and are entering another divergent stage.

Washing machines

Front- and top-loading automatic washing machines took about 40 years to become dominant designs and remain so, despite several recent innovations in clothes washing technology. It is to be seen whether machines using these new technologies will displace today's dominant designs.

Vacuum cleaners

Once established in the 1920s, the dominant upright and cylinder vacuum cleaner designs remained unchanged in basic technology until challenged in the 1990s by Dyson's bagless dual cyclone cleaner. As Gantz (2012) observed, in the 20 years before Dyson's disruptive innovation there were 'no earth shaking technologies or innovative visual designs to lift products above generic adequacy in a saturated market' (p. 155). The success of the cyclonic cleaner then stimulated the industry to develop a variety of bagless and cyclonic cleaners, followed by robotic and cordless designs.

Lamps and lighting

The incandescent light bulb emerged in the early twentieth century as the dominant design for domestic lighting and survived 50 years of competition from more efficient fluorescent, halogen, compact fluorescent and solid state LED lamps before being phased out in many countries. Solid state lamps are predicted to become the new dominant design by 2020.

Television

Electro-mechanical had largely been displaced by all-electronic television before the Second World War. Analogue CRT TVs became and remained the dominant design until the introduction in the early 2000s of digital television and TV sets with flat screens of increasing size and resolution which became the new dominant type. Hard disk recorders and Internet streaming have taken 30 years to become the current dominant technologies for recording or catching up on TV content. So although there are new dominant designs, innovation is continuing, for example towards ultra-high definition TVs with curved LED and OLED screens (Chapter 4, Figure 4.11).

Mobile phones

It took over 25 years for the touchscreen smartphone using 3G or 4G cellular communications to become the dominant design, having evolved from

> analogue to digital technology and from various handset designs made possible by increasingly powerful, efficient and compact electronic components. Despite the currently dominant touchscreen smartphone, innovation is continuing apace, for example towards 5G communications and 'wearable' products such as smartwatches.

The first lesson for business strategists, product planners, engineers, designers and marketers, drawn from the above sections, is that understanding patterns of innovation is important. This is because if they know where their business and products are located in the evolution of their industry, they should be better able to anticipate change, exploit new opportunities and avoid being overtaken by competitors or new disruptive technologies.

Follow new scientific knowledge and enabling technologies

A second conclusion from the case studies is that the products could not have been developed without prior scientific and technical knowledge and enabling technologies. The extent of this dependence varies for different products, with mechanical and electro-mechanical products, such as bicycles and vacuum cleaners, being based at least initially on craft knowledge and practical skills, while the electronic innovations of television and mobile phones could not have been invented without theoretical, scientific and technological foundations. However, as they evolve even craft-based products tend to depend for their continuing development on increasing inputs of scientific and technological knowledge and new technologies and materials. As well as providing the basis for new products, innovations in technology, materials and components can free designers to create new forms and types of product.

> ### Lamps and lighting
>
> The development of incandescent light bulbs in the late nineteenth and early twentieth centuries depended on several enabling technologies necessary to produce a practical and affordable product. These included a vacuum pump capable of removing most of the air from the bulb, the process to make drawn tungsten and technologies to manufacture light bulbs economically. In the late twentieth and early twenty-first century, the development of compact fluorescent lamps (CFLs), light emitting diodes (LEDs) and organic light emitting diodes (OLEDs) have enabled designers to create new forms of light fitting that utilise the compactness and cool operation of these devices (e.g. Chapter 3, Figures 3.8a, 3.8b and 3.11).

Vacuum cleaners

The development of plastics and injection moulding technology freed engineers and industrial designers to create new styles of vacuum cleaner that also cost less to manufacture. Subsequently new components such as compact and powerful digital motors, improved rechargeable batteries and computerised visual scanning systems have enabled the creation of innovative designs such as cordless and robotic cleaners.

Television

The scientific and technological foundations of television have been well expressed by Burns (1998) who wrote: 'all-electronic, high-definition television had to await the conclusions and the deductions of the ... sciences of photo-electricity, thermionic emission, electromagnetic wave propagation, electron optics, secondary electron emission, radio and electronics. Only then could the technology of television progress on a secure scientific basis' (pp. 611, 616).

Mobile phones

The shift from second (2G) to third generation (3G) mobile phone communications, in order to provide faster data transmission, depended on systems innovations, such as packet switching to replace circuit switching technology, as well as product innovations in handsets. Likewise, the shift from 3G to much faster 4G depends on use of an all-IP (Internet Protocol) network for both data and voice transmission. A modern smartphone could not have been designed without innovations in operating system and application software, longer-life batteries and strong, scratch-resistant screen glass.

To innovate successfully engineers, designers and manufacturers need to keep abreast of, and be capable of applying, new theoretical, scientific and technical knowledge and potentially relevant new technologies, materials, components and systems in both their own and related fields.

Designing for product success

While many innovations and new products fail to diffuse into widespread use, others have become highly successful in terms of adoption and/or commercial profitability. What do the case studies suggest distinguishes these successful innovations and new products from the less successful ones?

Genuine innovation

For a genuine innovation or 'first to the world' product to succeed it must offer a function or other benefit that previously did not exist and that consumers need or want.

Lamps and lighting

Solid state LED lamps (Chapter 3, Figure 3.7) offer improved efficiency, compactness, cooler operation and lower running costs over CFL and halogen incandescent lamps and so are gradually displacing these earlier technologies.

Vacuum cleaners

Dyson's original cyclonic cleaners offered several unique benefits that attracted many buyers. The benefits included no loss of suction as the machines filled with dust, no need to replace dust-bags, low dust emissions, and innovative-looking, user-centred designs. Despite the subsequent introduction of many rival products, Dyson has managed to compete through continuous innovation (e.g. Figure 5.13a), as well as brand recognition as the pioneer of the bagless cyclonic cleaner.

Television

Digital television offered the advantages of multiple channels, higher definition and better sound than analogue; and then flat panel LCD, LED and OLED screens have provided advantages of much slimmer TV sets with larger screens providing even higher picture quality and using less energy than previous CRT designs (Chapter 4, Figures 4.9b and 4.11).

Mobile phones

The iPhone when first introduced in 2007 offered many advantages over other smartphones, including greater ease of use, touchscreen icons and keyboard, an increasing number of apps and a highly desirable design. Other manufacturers soon developed their own touchscreen smartphones based on the concepts pioneered by the iconic iPhone (Chapter 6 Figure 6.10).

To succeed, a 'first to the world' product should offer potential consumers one or more unique functions or benefits that are not provided by existing products or services.

Relative advantage

Very few new products are 'first to the world' innovations; most are based on established technologies, designs and manufacturing processes and so must compete with rival products from other suppliers. To succeed, such new products must offer what consumers consider to be a genuine set of advantages over rival products, services or systems; what Rogers (1995) calls 'relative advantage'. For example, Sony's Trinitron colour TV tube offered better picture quality than conventional colour CRTs, something highly valued by consumers, and so made Sony TVs very successful in the 1970s (Chapter 4, Figure 4.3b). Conversely, the LaserDisc, a high-definition rival to videocassette recorders launched in the 1980s (Figure 7.3), failed the test of relative advantage. The player and LaserDiscs were more expensive than VCRs and videocassettes, could not be recorded on, and could only store a shorter recording. For consumers, all these disadvantages outweighed the LaserDisc's higher definition pictures.

One indicator of relative advantage is the assessment criteria used by *Which?* to help consumers choose between competing products. In general *Which?* recommendations are based on providing information to help consumers balance the specification and performance of a product, which may include its running costs and ease of use, against its price. A Best Buy or high-scoring product is one that, relative to other products, offers a desirable specification and excellent performance at a competitive price.

These rational criteria of product value to the consumer, however, do not take into account other factors that people use when choosing between products. These include brand image, reputation and loyalty; product style and fashionability; emotional appeal; marketing; online customer reviews; warranties; and practical issues such as availability or fitting available space in the home. Branding especially has been described as 'the most potent commercial and cultural force on the planet' (UEA, 2015). Designers and manufacturers can control the relative advantage offered by their products on some of these factors – for instance, Apple's highly valued brand and product aesthetics – but other consumer choice factors are not necessarily within their control.

FIGURE 7.3 Philips LaserDisc player (1982) failed to catch on because of its higher price and perceived lack of relative advantage over videocassette recorders

Washing machines

Current *Which?* recommendations, Best Buys and test scores for washing machines are based on price, specification (capacity, spin speed) and performance (cleaning and rinsing, ease of use, energy use and running cost) (Which? Ltd., 2014b). The criteria have evolved as washing machines have changed. For example, until 2005 reliability and servicing of individual models were important criteria, but as most machines have become more reliable these measures have been dropped in favour of brand reliability.

Vacuum cleaners

Current *Which?* recommendations, Best Buys and test scores for vacuum cleaners are based on price, specification (capacity, bagged/bagless, weight) and performance (cleaning of carpets, floors and pet hair; allergen retention; noise and reliability) (Which? Ltd., 2014b). The criteria have evolved as vacuums have changed. For example, pick-up of pet hair only began to be tested in 1999, after the introduction of cyclonic cleaners.

Lamps and lighting

Current *Which?* recommendations, Best Buys and test scores for light bulbs are based on price, specification (wattage, rated output, claimed life, fitting type); performance (actual output, warm-up time, efficiency, durability). The *Which?* criteria have evolved as lamps have developed from incandescent to fluorescent to LED technologies, but have always tried to balance price against performance in terms of light output, life and running cost, taking into account factors like shape, light quality, warm-up time and type of fitting.

Television

Current *Which?* recommendations, Best Buys and test scores for television sets are based on price, specification (screen size, features) and performance (picture and sound quality, ease of use and running cost) (Which? Ltd., 2014c). Again the criteria have evolved with TV technology. For example, with the increase in TV connectivity, built in wi-fi and video recording capability have become important parts of the specification.

Mobile phones

Current *Which?* recommendations, Best Buys and test scores for smartphones are based on price, specification (screen size, memory, operating system), performance (call quality, screen resolution, battery life, Internet speed,

> camera quality and ease of use) (Which? Ltd., 2014d). The criteria for assessing mobile phones have of course changed with mobile phone technology as new functions and features, such as cameras and Internet connectivity became available even on budget phones.

For their products to score highly in *Which?* tests and be chosen – at least by rational consumers – manufacturers must develop products that offer value via a set of advantages, especially an equal or higher specification and a better functional performance than rival products at a competitive price. They should of course pay attention to other important factors influencing consumer choice, especially branding.

Affordable price

When first introduced new products and innovations command premium prices and so are mainly adopted by wealthy consumers, enthusiasts and 'innovators' – people who like to acquire novel products when they first appear (Rogers, 1995). Then usually, with improved scale and efficiency of production, and in order to expand the market, manufacturers normally reduce prices. If the unique functions or advantages relative to the competition are considered by consumers to represent 'value for money', many more then begin to adopt or purchase the product.

This pricing pattern can be observed in all the product case studies. In some cases, such as washing machines and colour television in the 1970s, the affordability of new products also depended on available methods of financing ownership, such as loans, hire purchase, renting or usage contracts instead of cash payments.

> **Vacuum cleaners**
>
> Vacuum cleaners were originally aimed at addressing the 'servant problem' in wealthy households and only became more widely affordable in the 1950s. In 1993 Dyson was able to charge about twice the price of conventional products for his bagless cyclonic cleaner. As bagless and cyclonic designs from competitors became available, Dyson's company was still able to charge premium prices for its products as the return for continuing technical and design innovation.
>
> **Television**
>
> When first introduced in Britain in 1967 a colour TV set cost about £300 (about £4,750 today) so most people rented. Six years later the price had almost halved, sets had become more reliable so it became more worthwhile

> to buy, even with a loan. Today you can pay from £150 to nearly £3,000 for a high-definition TV, depending on screen size and resolution. However, the technology, product platforms and many components are common to the different models within and across the price ranges. Thus good quality TV sets have become affordable for almost everyone in industrialised countries, with up-market models available at premium prices for those who can afford or want them.

Manufacturers should aim to develop and make ranges, preferably based on common platforms or components, to compete on quality factors (especially specification and performance) in different price bands aimed at different consumer groups and markets.

Good design

In the early experimental phase of innovation, products are often conceived as assemblies of functional or engineering components with relatively little attention paid to their ease of use, form, colour, surface finish and user interfaces. As the products evolve, increasing effort is normally devoted to their industrial and ergonomic design in order to make the products more useable, visually and tactilely appealing, contemporary or fashionable.

In 1934 the industrial designer, Raymond Loewy, created evolution charts of the automobile, the railway coach, telephone, clock and other products. Loewy's charts reveal a general pattern in which the products all evolved towards more streamlined and simple forms (Cogdell, 2004). Since then industrial and product designers, and sometimes ergonomists, have been increasingly involved in designing and developing new products, especially in large companies. In small and medium-sized companies, too, the employment of industrial designers in product development teams has been shown to be associated with commercial success (Roy and Potter, 1993).

As noted earlier, *Which?* reports do not normally attempt to assess the appearance of the products it tests; instead the reports provide consumers with photographs of the products so that they can judge for themselves. However, most *Which?* reports rate the products in terms of ease of use.

> **Washing machines**
>
> Electric washing machines are a good example of a product that started as an assembly of functional parts – wooden tub, externally-mounted motor, drive chains or belts, wringer, etc. As they evolved the parts became more integrated; first with the mechanical parts enclosed, then with the cabinet

design changing from round tubs on legs reminiscent of earlier machines to box shapes (compare Chapter 2, Figures 2.3, 2.5, and 2.8a). With the introduction of the automatic washing machine the drying mechanism was integrated into the cabinet instead of being provided by an external wringer or separate spin drier. With further evolution washing machine controls became more sophisticated and so more attention was paid to user interface design and informative displays. The cabinets began to have slightly rounded forms and were sometimes offered in colours other than white. The latest machines are increasingly sleek in form with large flush glass portholes and electronic displays to echo contemporary product and kitchen aesthetics.

Vacuum cleaners

Vacuum cleaners are an outstanding example of how since the 1930s industrial designers have been employed to create more attractive, modern product styles, to help reduce production costs and make the products easier and more convenient to use (e.g. Chapter 5, Figures 5.6, 5.7a, 5.11a and 5.13b).

Television

Television set design involves a 50-year transition from an item of veneered wooden furniture with a small cathode ray tube screen and a few simple controls, via the bulky grey or black plastic cabinet widescreen TVs with remote controls (Chapter 4, Figure 4.8), to today's elegant TVs with slim, flat, almost borderless, screens of increasing size with remote controls or touch-screens to operate the set's smart functions (Figure 4.11).

Mobile phones

Portable mobile phone handset design has from the beginning involved input by industrial designers. Handsets have evolved from the bulky 'brick' phones of the 1970s (Chapter 6, Figure 6.3) via more compact and fashionable bar, flip and slide types with functions accessed by pushbutton and screen based interfaces (Figures 6.3 to 6.8) into slim and elegant multi-function touch-screen smartphones (Figures 6.10a and 6.10b). The hardware and software design and the shape, touch, finish and casing materials of rival phones are a key factor in their competitive success.

Input by industrial/product designers, and preferably ergonomists, from the early stages of the product development process is important to create successful products – which look good, are easy to use, and are designed as an integrated whole.

High reliability

One of the most important factors in consumer choice is product reliability. Brand reputation, guarantees and after-sales service are related factors in consumers' purchasing decisions. Manufacturers are fully aware of the importance of reliability and so most have ensured that their products have become more reliable over time, while at the same time repairs have generally become more difficult and costly.

> **Television**
>
> In the 1970s colour TV sets were not very reliable, on average requiring two service calls per year. Hence *Which?* advised consumers to buy with a service contract or rent (Consumers Association, 1972). Japanese brands and models were more reliable than European ones and their success in capturing much of the UK market was partly due to this high reliability. However, by the early twenty-first century TV sets with solid state electronics and flat panel screens had become very reliable and so *Which?* no longer included reliability in the criteria for rating individual models. Instead it reports on TV brands, giving each a reliability score depending on the percentage that are fault-free for up to five years (Which? Ltd., 2014e).

A reputation for product reliability, created by designing all products in a range to be as reliable as economically possible, is one of the most important factors in building a successful brand.

System compatibility

Many consumer products have to connect to or interface with other products and systems and so compatibility with these other technologies is essential. Electric lamps, for instance, clearly have to be compatible with different national supply voltages and types of light fitting. To diffuse widely the products have also to be compatible with consumer requirements and preferences and meet any prevailing national or international standards and legislation. Thus, the success of the first digital mobile phones was facilitated by the EU's agreement to adopt GSM digital technology, which became the standard most widely adopted outside the USA.

> **Washing machines**
>
> The importance of system interdependency and compatibility has been highlighted by a sociologist of technology, Elizabeth Shove (2003). She argues that clothes laundering should be viewed as a 'system of systems' in which washing machine manufacturers have to design their products taking into

> account the actions of detergent manufacturers, textile and fabric producers, households and users. This means, for example, that washing machines designed for different markets need to provide wash temperatures and programmes that suit the existing – and likely changes in – detergent formulations, clothing and laundry habits of consumers in different countries and climates.

Designing products for different countries, markets and consumers must take into account differences in technical standards, consumer preferences and regulations and also the products or services of other businesses and industries that are interdependent with, and might affect the development or use of, your products.

Avoiding product failure

Many new products and innovations fail to reach the market, or if they do, are not sold in sufficient numbers to provide a return on the investment in their development, or be adopted widely enough to be considered a success. What do the case studies suggest are the main reasons for product failure?

Premature innovation

The case studies provide several examples of new products or innovations based on technologies that were insufficiently developed to offer the performance or prices required to persuade enough consumers to adopt them. Premature attempts at innovation are found among 'first to the world' products and also among new products that offer insufficient relative advantages over existing products with a similar function. New or improved technologies, materials, manufacturing processes and designs, plus reductions in price, are usually necessary before such products are attractive enough to be widely adopted.

> **Lamps and lighting**
>
> Relatively few consumers bought early compact fluorescent lamps (CFLs) due to their cost, slow warm-up, dimness, size and weight compared to incandescent light bulbs. Instead most early domestic CFL adoptions were through subsidised energy-saving schemes. CFL sales only took off after price reductions and the development of more compact designs that offered faster warm-up and increased brightness. Nevertheless, CFLs only became commonplace after incandescent light bulbs began to be withdrawn from sale in the early 2000s. CFLs are now beginning to be displaced by more efficient and longer-lasting solid state LED lighting.

> **Television**
>
> Although personal battery-powered TVs were available in the 1980s, few were sold. It took 20 years for portable TV and video to become popular after the development of smartphones and tablet computers with 3G or wi-fi Internet access to online content.

Inventors, designers and manufacturers should avoid embarking on product development projects unless any required technologies, materials, components or production facilities are sufficiently developed to create an innovation that provides unique benefits or relative advantages compared to existing products or services and at a competitive price.

Inadequate complementary assets

A related aspect of premature innovation is that new products and innovations may depend for their operation on new external infrastructure networks or systems. In innovation theory these networks or systems are called 'complementary assets' (Teece, 1986). The new product or innovation cannot be widely adopted until that infrastructure has also been widely implemented; which takes time, money and often government approval.

> **Television**
>
> Adner (2012) gives an example of an innovation failure due to a company not having what he calls a 'wide lens' or 'innovation eco-systems' perspective, which included the complementary products and systems needed to make the innovation a success. Following Japanese high-definition television (HDTV) experiments, Philips Electronics developed analogue HDTV sets in the mid-1980s. But despite its advanced technology and excellent picture quality, Philips' HDTV failed because the HD cameras and transmission standards necessary for a high-definition television service had not yet been introduced. Philips did not benefit from its pioneering efforts when digital HDTV was finally commercialised 20 years later.
>
> **Mobile phones**
>
> Mobile phones depend on a radio communications network in order to function. The take-up of successive generations of mobile phones therefore depended on the rate of installation and degree of geographical coverage first of analogue 1G and then of digital 2G and 3G networks, as well as government approval of the radio frequencies to be used for cellular phones.

Product planners and marketers need a 'wide lens' or innovation eco-systems perspective before developing a new product or innovation to ensure that any complementary products and systems required to make it function have been developed and are sufficiently in place when the product is introduced.

Unnecessary 'bells and whistles'

As products evolve, there is a tendency for manufacturers to add new functions and features to differentiate their products from the competition and to persuade consumers to buy a new model. These functions and features sometimes include ones that consumers do not use, want or need.

> **Television**
>
> 3D TV sets entered the UK market in 2009. To see a high-definition 3D picture mostly required viewers to wear heavy battery-powered spectacles. (Although some 3D TVs only required lightweight 'passive' specs, this did not produce such a good 3D image.) 3D TV therefore was not a novel feature that persuaded many consumers to buy a new set, or stimulated providers to invest in making 3D content. Hence, although manufacturers continue to offer models with 3D, it is not widely used, especially given the limited 3D content available.

Manufacturers and designers should resist adding 'bells and whistles' to a product for marketing reasons which make the product more expensive or complex, unless they can identify a genuine consumer demand for the new function or feature.

High production costs

Products fail if they cannot be made profitably or in sufficient volume at a price that enough consumers can afford or are willing to pay. If making the product at an acceptable price requires investment in new equipment, factories or systems, the investment would have to be financially worthwhile if the product is to be a commercial success. In practice, it has meant that many consumer products can only be sold at a competitive price if they are manufactured in low-wage economies, mainly in Far Eastern countries.

168 Lessons for product designers, developers and innovators

Lamps and lighting

In the 1970s engineers proposed designs for compact fluorescent lamps (CFLs) to replace the incandescent light bulb. Although these designs often worked in the laboratory, putting them into mass-production was considered too expensive. It was not until Philips developed a CFL for mass-manufacture and made the necessary investments in new production equipment that CFLs could be launched onto the market. Even then, the costs of making these early CFLs meant they were sold at a relatively high price and this meant that they had a limited commercial market.

Vacuum cleaners

Despite being a strong advocate of manufacturing in Britain, in 2002 Dyson reluctantly moved production of his company's vacuum cleaners from Britain to Malaysia in order to reduce costs and to bring the factory closer to his component suppliers.

Successful innovation often requires investment in new production facilities or sourcing from low-cost manufacturers if the new product is to be sold profitably at an acceptable price.

Balance technology push and market pull

Disruptive innovations which lead to a whole new industry are rare and are normally the result of 'technology push'. Such innovations are created by visionary individuals or teams who perceive the potential for a genuinely new product or service to meet a latent consumer need that conventional market research cannot usually detect. To translate their idea, invention or conceptual design into a working product almost always requires considerable time, effort and investment, often in collaboration with other groups and organisations. The collaborators are typically other companies, government bodies, universities or research organisations that, if they are to produce a successful innovation, have teams of dedicated scientists, R&D engineers and designers and committed managers.

It is usually only after the innovation has been launched that 'market pull' begins to take effect. Consumer and market research can then help manufacturers decide how the new product can best be improved in response to customer demands and wishes.

> **Mobile phones**
>
> The first portable cell phone was the result of the vision of a Motorola engineer, Dr. Martin Cooper. It took Motorola's team of engineers and industrial designers over ten years to translate Cooper's vision into a working portable handset, the DynaTAC 8000X (Chapter 6, Figure 6.3). To realise the vision also required the US government to release part of the radio frequency spectrum to allow the first analogue cellular communications networks to be developed. Evolution of the mobile phone then involved a mix of the 'push' from innovations in micro-electronics, batteries, telecommunication technologies and networks and 'pulls' from consumer desires for more portable, better performing, easier to use and more fashionable designs.

Inventors and innovators require the vision to anticipate a latent demand for a genuinely new product for which few consumers will have perceived a need. After the innovation has been introduced, consumer and market research can help to guide product evolution. But to continue making improvements, the 'market pull of' consumer feedback and the 'technology push' opportunities offered by new technologies, materials, components and manufacturing processes should be considered together.

Designing for the environment

Making, transporting, using, maintaining and disposing of products all have impacts on the environment. Depending on the product, these impacts may include ecological damage (e.g. climate change, loss of landscapes and wildlife and water pollution); resource depletion (e.g. of non-renewable minerals and fresh water); and risks to human health (e.g. due to air pollution or toxic substances) (UNEP, 2005).

For many decades, the usual responses to such environmental problems were measures by manufacturers to reduce wastes and pollution *after* they had been produced; for example, by installing waste treatment plant or equipping cars with catalytic converters. However, from the late 1980s onwards some companies began to shift their attention from these 'end-of-pipe' approaches towards developing cleaner production processes, which generate less pollution and waste in the first place, and/or make more efficient use of energy and materials.

Then, with increasingly tough environmental legislation, standards and voluntary agreements, plus pressure from environmental groups and 'green' consumers and competition from 'greener' businesses, manufacturers began developing products with reduced environmental impacts as an important part of their design specification. Manufacturers therefore increasingly shifted their attention from trying to clean up their products' environmental impacts during manufacture, to designing out as many of the impacts as possible during product development.

Designing or redesigning products for the environment (DfE) may be undertaken to different levels using different approaches. Brezet (1997) proposed four levels of DfE, which have subsequently been termed Green design; Ecodesign; Sustainable design; and Sustainable innovation (Roy, 2006) (Figure 7.4).

Green design

Green design is the first level of DfE. It means focusing on one or two environmental objectives – for example, conserving resources by using recycled materials in the product or by improving its energy efficiency. Green design is the approach most manufacturers use when they begin to address a product's environmental impacts. They usually focus on the impacts they consider most important, are necessary to satisfy legislation or are easy to achieve, even if they are not the most significant.

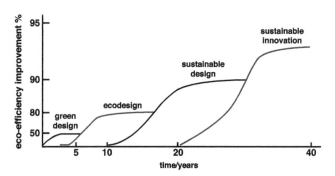

FIGURE 7.4 Different approaches to, or levels of, designing for the environment. (Adapted from Brezet (1997), p. 22 and Roy (2006), p. 90.)

Television

Apart from the issue of electronic waste from discarded sets, concern about the environmental impacts of television did not become a significant factor in equipment design until the early twenty-first century. However, there was growing concern about the amount of electricity used when TV equipment was left on standby. This stimulated regulatory actions, such as the 1999 International Energy Agency's *1 Watt Initiative*, which led to the average new TV set's standby consumption falling from about 5 watts to 1 watt or less.

Green design is a useful approach to DfE if there are one or more environmental impacts that clearly need addressing to satisfy legislation, meet consumer concerns or improve competitiveness. However, green design may fail to address a product's most significant impacts and so not provide the greatest potential for improving its environmental performance.

Ecodesign

Ecodesign is the second level of DfE. It is sometimes called life cycle design, because it attempts a balanced reduction in environmental impacts throughout a product's life cycle. Ecodesign does so by first assessing the product's various impacts from 'cradle to grave' and so enables product developers to focus on tackling the most important impacts.

In order to apply an ecodesign approach the various impacts (energy use, air pollution, toxic waste, etc.) of a product or product class should ideally be quantified using life cycle assessment (LCA). Quantifying what are the greatest impacts, and when they occur, helps to avoid focusing on minor environmental impacts while ignoring the major ones.

An important stimulus for manufacturers to use an ecodesign approach, informed by life cycle thinking and LCA studies, has been environmental policies and legislation. These include the EU's 2003 *Integrated Product Policy*, the 2009 *Ecodesign for Energy-related Products* Directive and Directives on Energy Labelling of products. This has resulted in a shift by large manufacturers, especially of energy-using products, from Green design to Ecodesign.

Washing machines

The introduction of EU environmental labelling required LCA studies to be carried out to inform policy-makers and manufacturers. The 1991 EU Ecolabel (Chapter 2, Figure 2.11b) for washing machines, for example, rested on a LCA study that demonstrated that over 90% of the environmental impacts of the machines occurred at the use phase of the life cycle. This determined that the main Ecolabel criteria for washing machines were low energy, water and detergent consumption. The EU Energy Label (Figure 2.12a), which replaced the washing machine Ecolabel, also rates products according to their energy and water consumption.

Mobile phones

LCA studies of mobile phones carried out by different companies demonstrated that the impacts of their products vary widely, but were concentrated

> on the raw materials extraction, component manufacture and use phases of the life cycle. Hence different companies focused their efforts on different measures, from Apple auditing its Far Eastern factories to ensure pollution compliance to Nokia designing energy-efficient phone chargers. Subsequently the companies have begun to use ecodesign approaches to assess and address their products' impacts that occur at different life cycle stages.

A major drawback of ecodesign informed by LCA is that it is complex and expensive and so cannot be carried out for every product. The information that LCA provides is also complex and often difficult to translate into a design specification or a design. Most companies therefore employ simplified ecodesign methods based on life cycle *thinking*, rather than formal LCA studies. One such method developed by Philips is called Green Focal Areas.

At Philips, product development teams are expected to focus on reducing impacts in one or more of six Green Focal Areas – Energy efficiency; Packaging; Substances; Weight; Recycling and disposal; and Lifetime reliability – when creating new or improved products (Philips, 2014). Philips adopted this simplified life cycle approach after its divisions had found LCA-based ecodesign methods too complex. Philips found that using this life cycle thinking checklist helps to identify the most important impacts of products. For example, as well as energy efficiency, product life is important for lighting products because a longer life saves both costs and materials.

Ecodesign approaches help to ensure that the main impacts of a product are addressed during product development. As a guideline, for mains-powered products the energy impacts during operational use almost always dominate. For battery-powered products, materials extraction, manufacturing of components and batteries, and product use can all give rise to significant impacts. For unpowered products (e.g. bicycles, furniture) the materials and manufacturing phases normally dominate. A simplified life cycle thinking approach such as Philips' Green Focal Areas is often a more practical method for carrying out ecodesign.

Sustainable design

Sustainable design is the third DfE level. It aims to provide the essential function of a product using the least environmentally harmful technical solution, for example, by using solar energy instead of grid electricity or batteries to power a product. Sustainable design also includes social and economic considerations, such as a product's fair trade implications or the health and safety of the workers making it. Sustainable design thus attempts to satisfy what have been called the 'three pillars of sustainability' – people, profit and planet (Crul *et al*, 2009).

Mobile phones

The clearest example of practical attempts to implement sustainable design concerns the development and manufacture of smartphones. Major suppliers such as Microsoft and Apple attempt to reduce the energy and resource use of their smartphones by incorporating power-saving software and minimising the amount of material used, and to eliminate more harmful substances than required by legislation. They have also been persuaded by pressure groups and international regulations to consider materials sources and factory conditions when designing and making their phones, for example to avoid 'conflict minerals' and child labour. An 'ethical' smartphone specifically designed for sustainability called Fairphone has been developed and was launched in 2013 by a Dutch social enterprise.

Datschefski (2001) has proposed the following checklist of five principles to help product developers apply a sustainable design approach. *Cyclic* – the product is made from natural materials or from minerals that are recycled in a closed loop. *Solar* – the product in manufacture and operation uses only renewable energy. *Safe* – all releases to air, water or land are 'food' for other systems. *Efficient* – requires up to 90% less energy, materials and water than equivalent products of ten years ago. *Social* – its manufacture and use supports basic human rights and natural justice. Such an approach will normally require collaboration between the many stakeholders involved in developing and introducing the sustainable product.

Sustainable innovation

Sustainable innovation, the highest DfE level, is even broader in scope than sustainable design and goes beyond technical solutions. Sustainable innovation involves providing a particular function (such as illumination or home entertainment) using environmentally optimal product–service mixes or socio-technical systems. It also requires developers to take into account the social and economic sustainability of any proposed new product–service or system.

Washing machines

The 'future of laundry' section in Chapter 2 detailed some of the innovations and concepts that in combination could provide a more sustainable system for cleaning clothes. For example, the new system might include:

- Innovative washing machines (e.g. a bead washer that uses very little water)
- Communal or commercial laundries equipped with the most environmentally efficient technologies (e.g. heat recovery and water recycling as well as innovative washing machines)
- A local clothes collection and delivery service.

Sustainable innovation should be viewed as part of long-term changes in society and, like sustainable design, requires collaboration between many stakeholders likely to be involved in planning, developing and implementing the new product–service or socio-technical system.

The rebound effect

The above DfE approaches aim to reduce the environmental impacts of products and systems – a so-called *eco-efficiency* strategy. This strategy, however, largely ignores the effects of people's behaviour and social changes, which may reduce the effectiveness of eco-efficiency measures. For example, consumers may choose to buy larger or more powerful products, which use more resources to make or energy to operate; or they may buy several products instead of one, or use them more frequently. Social change, such as the increasing number of single-person households, also has the effect of increasing the amounts of domestic equipment bought and used.

The case studies offer several examples. Consumers are choosing large-screen, smart TVs that use more energy than smaller, non-smart models and to own several sets per household. Consumers are installing additional lighting in their homes and tend to leave lights on longer after fitting low-energy light bulbs (Herring and Sorrell, 2008). Many consumers replace their mobile phones every one or two years despite having a fully functioning older model.

These are examples of what has been called the direct 'rebound effect'. The evidence for the size of the rebound effect is patchy, but for domestic appliances and lighting has been measured to reduce energy efficiency improvements by up to 12% (Greening *et al*, 2000). However, in newly industrialised countries where goods such as electric lamps and washing machines may be owned for the first time, the rebound effect can be considerably greater (Herring and Sorrell, 2008).

Even though the direct rebound effect may be relatively small, at least in industrialised countries, designers and manufacturers can try to reduce it further. Possible measures include providing consumers with information on how to operate products efficiently and designing products with economy or energy-saving settings. In order to mitigate the impacts of increased consumption, products can be designed for repair or upgrading and/or designed for reuse or easy recycling. It may also be worth developing products that meet the needs of the increasing number of single-person households.

When designing products for the environment, the direct rebound effect of increased consumption should be considered and, if possible, measures taken to mitigate it.

Understand social influences and impacts

Inventors, engineers, designers and manufacturers need to understand that new products and innovations are not the result of inevitable scientific and technological progress or human creativity. Socio-economic, political, commercial and cultural forces influence, or sometimes determine, which new products and innovations are created, introduced and adopted, and how they are used.

An understanding of the social forces shaping technology has developed from both practical experience and academic research. For example, social construction of technology (SCOT) theory argues that innovations do not evolve through the linear progress of science and technology. Instead technology is 'shaped' by the social, economic and political interactions and interests of people and institutions (relevant social groups or 'actors') in society (Bijker, 1995). Social studies of science and technology is another relevant strand of academic research. For example, Shove (2003) argues that people's behaviour in the use of products and technologies is strongly influenced by changing conceptions of what is 'normal', itself shaped by powerful cultural and economic forces.

Washing machines

As part of post-war cultural shifts in Western society towards higher standards of cleanliness and hygiene, and as more textiles became machine washable, clothes and linen began to be washed more frequently. The spread of automatic washing machines, with or without a tumble drier, on average has increased the weekly wash to an almost daily wash in industrialised countries (Shove, 2003).

Vacuum cleaners

Like washing machines, the design and marketing of vacuum cleaners has, since their invention, been shaped by social and cultural concerns about cleanliness and hygiene in the home. The latest manifestation is the development of 'animal' models specially designed to pick up pet hairs and the use of high-efficiency filters or multiple cyclones for removing the finest particles and allergens from the cleaner's exhaust air (e.g. Chapter 5, Figure 5.13a).

> ### Lamps and lighting
>
> An example of the social construction of technology is the 'Phoebus cartel'. In 1924 several major lighting manufacturers agreed to standardise the average life of an incandescent light bulb at 1,000 hours, although bulbs lasting longer had previously been made. They also agreed to avoid competing with each other, in order to increase their sales and profits (Peretti, 2014). This artificial constraint on lamp performance by vested commercial interests had to be abandoned due to the demands of the Second World War.
>
> ### Television
>
> Burns (1998) describes how the historical development of television was not simply the result of a series of scientific and technological innovations. Its evolution was shaped by government controls on broadcasting frequencies, international rivalries and alliances on broadcast standards, patent disputes, and many other non-technical factors.

Although inventors, designers and engineers cannot control social, cultural, commercial and political forces, it is important that they appreciate that such forces usually affect innovation. Innovators therefore need to take into account the wider context in which they are attempting to innovate.

Designing for the future

Anticipating future technologies, or classes of consumer product, is difficult. But it is possible to discern some general trends. And, of course, some projects now at the R&D stage will result in the products and innovations of the future.

Computers in product development and manufacture

Computers have become essential tools for computer-aided design and manufacture and rapid prototyping, as well as for remote collaboration between research engineers, product development teams and companies around the world. Computer communications have enabled the globalisation of production, and computer-aided design, rapid prototyping and 3D printing have allowed companies to produce a much greater variety of products and variants than was previously possible. This has provided consumers with a huge increase in product choices as well as customised products for different markets. But it has also encouraged companies to launch new models more frequently, leading consumers to discard and replace products more often. The shorter product development and replacement cycles are resulting in greater resource consumption and waste (Peretti, 2014).

ICT in consumer products

Information and communications technology (ICT) is being applied in many more consumer products. This goes beyond the information processing capabilities of high-definition TVs, 'smart' controls on washing machines and robotic vacuum cleaners. For example, washing machines connected to smart meters or the Internet will enable the machine to be turned on remotely by an energy supplier when the grid is lightly loaded or when a domestic solar energy system is generating electricity (Bourgeois *et al*, 2014).

This is part of a general trend towards the interconnection of different products and systems. Examples include mobile phones becoming the remote control interface for lights, washing machines and televisions and TVs sharing content wirelessly with other devices such as tablet computers and mobile phones. As part of this process, high technology companies such as Apple, Microsoft and Amazon try to get consumers to buy into their whole 'eco-systems' of related products and services with smartphones as the integrating component of the system. Google's ecosystem, for example, attempts to encourage consumers not only to use its Internet browser and search engine, but also buy its tablet computers and smartphones with its Android operating system, use its email, online office and software services and buy its apps (Which? Ltd., 2015).

User-centred and inclusive design

Manufacturers and designers are beginning to understand the importance of developing products to meet the needs, wants and demands of customers and users that go beyond what can be discovered by conventional market research. User-centred design involves, for example, making detailed observations of user behaviour when using products and systems; product developers collaborating with consumers in creative workshops; and designers becoming users in order to understand how to improve the consumer experience.

Inclusive design extends the user-centred approach by trying to design for the needs and abilities of all potential users, including the young, the old and people with disabilities. The idea is that if a product is specified for those who may have difficulties, it should be easy to use by all – for example, by ensuring that the controls and displays of a washing machine or smartphone can be understood and operated by anyone.

Donald Norman (2007) has pointed out, however, that the simultaneous trends towards making products more user-centred and automating them can conflict. He gives the examples of the built-in programmes on products such as washing machines, microwave ovens and heating controls that unnecessarily restrict what the user is able to do. Norman therefore advocates designing domestic equipment that provides for appropriate inputs of human and machine intelligence in their operation and maintenance.

Sustainable design and innovation

Designing for the environment and sustainability has become increasingly important, not just for the sake of the planet but to satisfy the wishes of many consumers, the needs of producers and for profit too. Thus, all the major manufacturers of lighting, washing machines, vacuum cleaners, TV equipment and mobile phones now have detailed environmental and sustainability policies and action plans – it is only necessary to search the companies' websites for details.

Designing for the environment is evolving from narrow Green product design approaches to designing more sustainable product–service mixes and socio-technical systems innovations. Many more examples of the latter sustainable approaches are being tested or introduced. For instance, it is possible to lease LED lighting as a service package, thus overcoming the upfront cost of purchasing an efficient lighting system (Chapter 3). And there are trials of clothes leasing services that provide clothes for rent or loan and which also wash, clean, iron and repair them (Chapter 2).

The shift to sustainable design and innovation is of growing urgency as demand for and ownership of consumer products spreads from industrialised countries, first to newly industrialised countries such as China, India and Brazil and then, with increased electrification, to low-income countries. For example, in the developing world 2 billion people already own a washing machine and an estimated further 3 billion will want one by 2050; nearly 70% of households in the developing world have a TV set but so far only a third of African households do, and the level of smartphone ownership in China already exceeds that in Britain (ITU, 2013). In the future global ownership of these products seems likely almost to reach saturation, which, combined with the expected increase in population to 9.6 billion by 2050, will produce unsustainable emissions and pollution levels and pressure on natural resources unless future products, services and systems are designed for sustainability.

Product planners, engineers, designers and manufacturers need to be aware of the broad trends, developments and experiments that are shaping the future of products, product services and socio-technical systems and build these likely changes into their thinking about what is possible and what could be worth developing for a more sustainable future.

References

Abernathy, W. J. (1978) *The productivity dilemma*, Baltimore, MD, Johns Hopkins University Press.

Adner, R. (2012) *The Wide Lens. A New Strategy for Innovation*, New York, Penguin.

Bijker, W. E. (1995) *Of Bicycles, Bakelites, and Bulbs: Toward a Theory of Sociotechnical Change*, Cambridge MA, MIT Press.

Bourgeois, J., van der Linden, J., Kortuem, G. and Price, B. (2014) 'Conversations with my washing machine: interactive energy demand-shifting with self-generated energy', paper

presented at the Energy Research Conference, Milton Keynes, The Open University, 3 April.
Brezet, H. (1997) 'Dynamics in ecodesign practice', *UNEP Industry and Environment*, vol. 20, nos 1–2, January–June, pp. 21–24.
Brezet, H. (1998) *Ecodesign: A Promising Approach to Sustainable Production and Consumption*, Paris, United Nations Publications.
Burns R.W. (1998) *Television: an International History of the Formative Years*, London, Institution of Electrical Engineers.
Christensen, C. M. (2000) *The Innovator's Dilemma*, New York, First Harper Business.
Cogdell, C. (2004) *Eugenic Design: Streamlining America in the 1930s*, Philadelphia, University of Pennsylvania Press.
Consumers Association (1972) 'Colour TV rentals, reliability and servicing', *Which?* September, pp. 279–286.
Cooper, R.G. (2011) *Winning at New Products* (4th edn), New York, Basic Books.
Crul, M., Diehl, J. and Ryan, C. (2009) *Design for Sustainability: a Step-by-step Approach*, Paris, United Nations Environment Programme.
Datschefski, E. (2001) *The Total Beauty of Sustainable Products*, Crans-près-Céligny, Switzerland, RotoVision.
Dosi, G. (1982) 'Technological paradigms and technological trajectories: A suggested interpretation of the determinants and directions of technical change', *Research Policy* vol. 11, pp. 147–162.
Foster, R. N. (1988) *Innovation: the Attacker's Advantage*, New York, Summit Books.
Gantz, C. (2012) *The Vacuum Cleaner: A History*, Jefferson, North Carolina and London, MacFarland and Co.
Greening L.A., Greene, D.L. and Difiglio, C. (2000) 'Energy efficiency and consumption – the rebound effect – a survey', *Energy Policy*, vol. 28, nos 6–7, pp. 389–401.
Herring, H. and Sorrell, S. (eds) (2008) *Energy Efficiency and Sustainable Consumption. The Rebound Effect*, London, Palgrave Macmillan.
ITU (2013) *Measuring the Information Society*, Geneva, International Telecommunication Union.
Kotler, P. (2000) *Marketing Management* (Millennium edition), New Jersey, Prentice-Hall.
Norman, D. A. (2007) *The Design of Future Things*, New York, Basic Books.
Peretti, J. (2014) *The men who made us spend*. Episode 1, BBC2 TV, 28 June.
Philips (2014) Our Green Products [Online]. Available at http://www.philips.com/about/sustainability/ourenvironmentalapproach/greenproducts/index.page (accessed February 2015).
Porter, M. E. (1990) *The Competitive Advantage of Nations*, New York, The Free Press.
Rogers, E.M. (1995) *Diffusion of Innovations* (4th edn), New York, The Free Press.
Roy, R. and Potter, S. (1993) 'The commercial impacts of investment in design', *Design Studies*, vol. 14, no. 2, pp. 171–193.
Roy, R. (2006) 'Products: New product development and sustainable design', Block 3, T307 *Innovation: Designing for a Sustainable Future*, Milton Keynes, The Open University.
Roy, R. (2013) Creative design, Book 3, T217 *Design Essentials*, Milton Keynes, The Open University.
Shove, E. (2003) *Comfort, Cleanliness and Convenience: The Social Organization of Normality*, Oxford, Berg.
Teece, D.J. (1986) 'Profiting from technological innovation: Implications for integration, collaboration, licensing and public policy', *Research Policy*, vol. 15, no. 6, pp. 285–305.
Tidd, J. and Bessant, J. (2013) *Managing Innovation: Integrating Technological, Market and Organisational Change* (5th edn), Chichester, Wiley.

UEA (2015) *The Secret Power of Brands*, Future Learn course, Norwich, University of East Anglia.

UNEP and Delft University of Technology (2005) Design for Sustainability: A Practical Approach for Developing Economies [Online]. Available at http://www.d4s-de.org/manual/d4stotalmanual.pdf (accessed September 2014).

Utterback, J. M. and Abernathy, W. J. (1975) 'A dynamic model of product and process innovation', *Omega*, vol. 3, no. 6, pp. 639–656.

Utterback, J.M. (1994) *Mastering the Dynamics of Innovation*, Boston, MA, Harvard Business School Press.

Which? Ltd. (2014a) 'Is it time to upgrade to a newer TV?' *Which?* September, pp. 64–66.

Which? Ltd. (2014b) 'Test lab buyers guide. Laundry and cleaning appliances', *Which?* September, pp. 76–81.

Which? Ltd. (2014c) 'Test lab buyers guide. TVs, audio, in-car and home entertainment', *Which?* August, pp. 68–73.

Which? Ltd. (2014d) 'Test lab buyers guide. Computers, cameras and phones', *Which?* June, pp. 76–81.

Which? Ltd. (2014e) 'Best tech brands 2014', *Which?* July, pp. 20–25.

Which? Ltd. (2015) 'Technology eco-systems explained', *Which?* January, pp. 30–32.

ILLUSTRATION ACKNOWLEDGEMENTS

Grateful acknowledgement is made to the following sources for permission to reproduce material within this book:

1.1	Which? Ltd.
1.2, 1.3, 1.4, 1.6, 1.7, 1.8, 1.9 and 1.11	Robin Roy
1.5a and 1.5b	© Moulton Bicycle Company
1.10	Robin Roy courtesy of Matthew Cook
2.1a and 2.1b	http://www.oldandinteresting.com/history-washing-machines.aspx
2.2b, 2.7 and 2.11a	Robin Roy
2.2a, 2.3, 2.4, 2.5 and 2.6b	www.oldewash.com courtesy of Lee Maxwell
2.6a	United States Patent Office US Patent 2,150,638, 1939
2.8a, 2.8b, 2.8c, 2.8d and 2.9	Which? Ltd.
2.10	Adapted from Roy, R. (2006) 'Products: new product development and sustainable design', T307 *Innovation* Block 3, Milton Keynes, The Open University, p.132 and Durrant, H.E., Hemming, C.R., Lenel, U.R. and Moody, G.C. (1991) *Environmental labelling of washing machines. A pilot study for the DTI/DOE*, Cambridge, UK, PA Consulting Group, August, Figure 2.3 p.37.
2.11b	http://ec.europa.eu/environment/ecolabel/ © European Union 1995-2015

Illustration acknowledgements

2.12	http://www.newenergylabel.com/uk/labelcontent/washers European Committee of Domestic Equipment Manufacturers © CECED 2015
2.13a	Samsung
2.13b	Dyson Ltd.
2.14	Xeros Ltd.
2.15	Courtesy of Emma Dewberry from a draft of a paper by Dewberry, E., Cook, M., Angus, A., Gottberg, A., and Longhurst, P., 'Critical reflections on designing product service systems', Figure 3.
3.1a, 3.1b, 3.3a, 3.3b, 3.4, 3.5, 3.6b, 3.7, 3.8a, 3.8b and 3.10	Robin Roy
3.2	http://www.archives.gov/education/lessons/telephone-light-patents/#documents Thomas Edison's Electric Lamp Patent Drawing. Record Group 241 Records of the Patent and Trademark Office National Archives and Records Administration. National Archives Identifier: 302053
3.6a	Echtner / http://en.wikipedia.org/wiki/Compact_fluorescent_lamp#/media/File:Old_compact_fluorescent_lamp.JPG This file is licensed under the Creative Commons Attribution-Share Alike http://creativecommons.org/licenses/by-sa/3.0/deed.en
3.9	Adapted from US DOE (2013a) Life-Cycle Assessment of Energy and Environmental Impacts of LED Lighting Products [Online]. US Department of Energy, Building Technologies Office, April, http://apps1.eere.energy.gov/buildings/publications/pdfs/ssl/lca_factsheet_apr2013.pdf p.2
3.11	https://www.flickr.com/photos/philips_newscenter/7096089207/in/set-72157626859053740 © Philips Communications, 2012
4.1, 4.2, 4.3a, 4.6, 4.7a and 4.7b	http://www.earlytelevision.org/ Early Television Museum Courtesy of Steve McVoy
4.3b	Rob Skitmore / http://objectwiki.sciencemuseum.org.uk/wiki/Image_E2005.343.2_Sony_Trinitron_TV.html This file is licensed under the Creative Commons Attribution Licence http://creativecommons.org/licenses/by/2.0/
4.4a	Groink / http://en.wikipedia.org/wiki/File:JVC-HR-3300U.jpg This file is licensed under the Creative Commons Attribution-Share Alike Licence http://creativecommons.org/licenses/by-sa/3.0/
4.4b	http://de.wikipedia.org/wiki/Betamax This work has been released into the public domain by its author Battenburg at the German Wikipedia project
4.5, 4.7c, 4.8, 4.9a, 4.9b, 4.10a and 4.10b	Robin Roy

Illustration acknowledgements

4.11	Samsung
5.1, 5.4, 5.7b, 5.9, 5.10a, 5.10b and 5.12	Robin Roy
5.2	Taken from https://agentinfidel91321.wordpress.com/2014/04/16/the-motorized-vacuum-cleaner/
	US Patent US889823 (1908). Taken from http://www.google.com/patents/US889823
5.3a	http://en.wikipedia.org/wiki/Vacuum_cleaner#/media/File:NMAH_DC_-_IMG_8859.JPG This work has been released into the public domain by its author Daderot
5.5	Taken from http://group.electrolux.com/en/history-1920-1929-737/
5.6	ttp://commons.wikimedia.org/wiki/File:Electrolux_vacuum_cleaner_Model_30_DMA.jpg This work has been released into the public domain by its author User FA2010
5.7a	INTV1980 / http://en.wikipedia.org/wiki/File:Hoover_model_29_ad.jpg This file is licensed under the Creative Commons Attribution-Share Alike Licence http://creativecommons.org/licenses/by-sa/3.0/
5.8a, 5.8b, 5.8c and 5.8d	Which? Ltd.
5.11a, 5.11b, 5.13a and 5.14	Dyson Ltd.
5.13b	Miele GB
6.1, 6.2, 6.7a, 6.7b, 6.9, 6.10a	Robin Roy
6.3	Redrum0486 / http://en.wikipedia.org/wiki/File:DynaTAC8000X.jpg/. This file is licensed under the Creative Commons Attribution-Share Alike Licence http://creativecommons.org/licenses/by-sa/3.0/
6.4	Jkbw / http://en.wikipedia.org/wiki/File:Nokia_1011.jpg. This file is licensed under the Creative Commons Attribution-Share Alike Licence http://creativecommons.org/licenses/by-sa/3.0/
6.5	Nkp911m500 /.http://commons.wikimedia.org/wiki/File:First_Generation_Motorola_StarTAC_cellular_phone.jpg This file is licensed under the Creative Commons Attribution-Share Alike Licence http://creativecommons.org/licenses/by-sa/3.0/
6.6	krystof.k / http://commons.wikimedia.org/wiki/File:Matrixphone.jpg?uselang=de This file is licensed under the Creative Commons Attribution-Share Alike Licence http://creativecommons.org/licenses/by-sa/3.0/
6.8	http://en.wikipedia.org/wiki/Motorola_Razr#/media/File:RAZR_V3i_opened.JPG This work has been released into the public domain by its author Peterwhy

6.10b	Samsung
6.11	Karlis Dambrans / http://en.wikipedia.org/wiki/File:Blackberry_Q10_home_screen.jpg This file is licensed under the Creative Commons Attribution-Share Alike Licence http://creativecommons.org/licenses/by-sa/2.0/
6.12	ZolHaj / http://en.wikipedia.org/wiki/File:Nokia_100,_front_and_back.JPG. This file is licensed under the Creative Commons Attribution-Share Alike Licence http://creativecommons.org/licenses/by-sa/3.0/
6.13	From Muir Wood, A. (2010) *The Nature of Change in Product Design. Integrating Aesthetic and Technical Perspectives*, PhD thesis, Design Management Group, Institute for Manufacturing, Cambridge, University of Cambridge, September, Figure 7.13 p.131. Courtesy of Andrew Muir Wood.
6.14	http://en.wikipedia.org/wiki/Sony_Ericsson_T610#/media/File:SonyEricsson_T610_AluminumHaze_front.jpg This work has been released into the public domain by its author, Episteme.
6.15	Courtesy of The Gaia Foundation, www.gaiafoundation.org
6.16	Loïc Le Meur / http://en.wikipedia.org/wiki/Google_Glass#/media/File:A_Google_Glass_wearer.jpg 'A Google Glass wearer' – Flickr: Loïc Le Meur on Google Glass. This file is licensed under Creative Commons Attribution 2.0 Generic http://creativecommons.org/licenses/by/2.0/
7.1	Adapted from Utterback, J. M. and Abernathy, W. J. (1975) 'A dynamic model of product and process innovation', *Omega* vol. 3, no. 6, p .645 and Abernathy, W. J. (1978) *The productivity dilemma*, Baltimore, MD, Johns Hopkins University Press, p. 72.
7.2	Adapted from Utterback, J.M. (1994) *Mastering the Dynamics of Innovation*, Boston, MA, Harvard Business School Press p. 160.
7.3	Robin Roy
7.4	Adapted from Brezet, H. (1997) 'Dynamics in ecodesign practice', *UNEP Industry and Environment*, vol. 20, nos. 1–2, January–June, p. 22 and Roy, R. (2006) 'Products: New product development and sustainable design', Block 3, T307 *Innovation: Designing for a Sustainable Future*, Milton Keynes, The Open University, p. 90.

INDEX

Ahlgren, Erik 132
Air Recycling vacuum cleaners 116
all-electronic television systems *see* CRTs (cathode ray tube) televisions
analogies 5, 6, 154
analogue television 76–7
Android operating system 124, 137, 177
Apple: environmental issues 135, 136; ethical issues 136–7; iPhone 129–30, *130*; smartphones 123, 124, 125, 158
apps 83, 119, 124, 129–30
arc lamps 48
ARM microchips 151
AT&T 72, 120
Australia 27, 30, 63, 97

bagless vacuum cleaners *see* cyclonic vacuum cleaners
Baird, John Logie 72, 73
Baird televisions *73*, *74*, 83, *85*
Batteries Directive 134
BBC (British Broadcasting Corporation) 72, 73, 74, 76, 77, 79, 82
Beatty Bros washing machine *24*
Beetham, Edward 22, *23*
Bell Laboratories 72, 75, 119, 120
Bell System 120
Bendix washing machines 25, *26*, *28*, 38
Best Buy evaluation 3, *3*, 159, 160–1
bicycle case study 5–14; design phases 5–13; environmental impacts 13; social influences and impacts 13–14; *Which?* tests 9, 12

Bissell carpet sweepers 98, *99*, 100
Black and Decker vacuum cleaners 107, *107*, 113–14
Blackberry mobile phones 124, 131, *131*, 140
Blu-ray 81, 82, 92
Booth, Hubert 98, *99*
Bosch washing machines 34, *34*
branding 159, 161
Brazil 22, 39, 97
British Broadcasting Corporation (BBC) 72, 73, 74, 76, 77, 79, 82
British Telecom 79, 121
Burrows, Mike 11, *11*
Bush televisions 84, *85*

cable television 79, 81
camcorders 77, 78
Candy washing machines *29*
canister vacuum cleaners *see* tank vacuum cleaners
car phones 119, 120, 125, *125*
carbon filament lamps 48–50, *49*
carpet sweepers 98, *99*, 100
cathode ray tube televisions (CRTs) *see* CRTs (cathode ray tube) televisions
CBS television system 75
Cellnet 121
cellular networks 120–1
child labour 136–7
China 12, 97, 119, 139, 178
circular economy 137
clothes ownership 42–4

cold water washing 30, 39
colour televisions 72, 75–6, *75*, 77, 77, 80–1
compact fluorescent lamps: environmental impacts **61**, 62–3, *62*, 63, *63*; premature innovation 165; production costs 168; technology 56–7, *57*, 59, *60*
complementary assets 166–7
computers in product development 176
conflict minerals 136, 137, 173
consumer organisations 17
Consumers' Association *see Which?* magazine
Cooper, Martin 125–6, 169
cordless vacuum cleaners 113–14
creative ideas 154
CRTs (cathode ray tube) televisions: colour 72, 75–6, *75*, 77, 77, 80–1; design 83, 84, *84*, *86*, 150; early systems 72, 73–4, *74*
CT2 phone point 126–7
curved screen televisions *92*, 93
cyclonic vacuum cleaners: case study 108–11, *109*, *110*, *111*, *113*; disruptive innovations 107, 112; environmental impacts 114; prices 109, 110; production costs 168; robot 116, *116*
cylinder vacuum cleaners *see* tank vacuum cleaners

design change theories 14–17
design classics 8, 16, 100, 106; *see also* fashionable designs
design for all 177
design phases *see* dominant design phases; experimental design phases; innovative design phases
designing for the environment (DfE) 169–75, *170*
detergent-free washing machines 35, 39–40
digital mobile phones 127–32
digital television 79–81, 82–3, 85, 158
disruptive innovations 17, 107–12, 168–9
dominant design phases 14, 15–16, 149–51; bicycles 8–9; lamps 52–3; mobile phones 124, 131–2, 142, 151; television 76–7, 84, 93, 150; vacuum cleaners 100–7; washing machines 25–9, 150
Dreyfuss, Henry 103, *105*
Dunlop, John 7–8
durability 52, 62, 115, 153, 160
dust bags 106–7, 108, 114

DVD products 80, 82, 89
Dyson: vacuum cleaners 108–11, *109*, *110*, *111*, *113*, *116*, 168; washing machines 35, *36*
Dyson, James 107, 108

ease of use assessments 162
eco rating of mobile phones 135–6
ecodesign 16, 63, 89, 114–15, 171–2
Ecolabels 29–30, 31–3, *33*, 171; *see also* Energy Labels; Energy Star
ecological rucksacks *see* life cycle assessments
Edison, Thomas 48–50, *49*, 52
Electric Suction Sweeper Company 100, *102*
electrical and electronic waste 90, 133
electricity generation and distribution 50
electricity use 22, 60, 64, 65, 87–9
electro-mechanical television systems 72–3, *73*, 75–6
Electrolux: vacuum cleaners 100, 101, 103, *103*, *106*, 112–13; washing machine trial 41
electronic programme guides 86
electronic television systems *see* CRTs (cathode ray tube) televisions
EMI televisions 73
enabling technologies 156–7
end-of-life reuse 133, 137, 174
energy efficiency: lamps 52–3, 54, 56–7, 60–1, **61**, 153; televisions 88–9
Energy Labels: lamps 63; televisions 89, *89*; vacuum cleaners 114; washing machines 33, 34, *34*, 171; *see also* Ecolabels; Energy Star
Energy-related Products (ErP) Directive 89, 171
Energy Star 34, 89; *see also* Ecolabels; Energy Labels
English Electric washing machines *28*
environmental impacts 16, 150, 169–75; bicycles 13; lamps 60–3, **61**; mobile phones 132–6; televisions 87–90; washing machines 29–30, *30*, 31, 37, 38–9
environmental labelling *see* Ecolabels; Energy Labels; Energy Star
ethical issues, mobile phones 136–7
EU Directives 63, 89, 90, 133–4, 171; *see also* Energy Labels
experimental design phases 5–8, 14–15, 22–4, 72–4, 83, 119–21, 148–9

Index **187**

failure avoidance 80, 165–8
Fairphone ethical smartphone 137
Farnsworth Television 73
fashionable designs 15–16, 162; bicycles 9, 12; mobile phones 122, 127, 129, *129*, 151, 163, 169; televisions 83; *see also* design classics
features, unnecessary 167
Fingerworks 129
Fisher, Alva J. 24
5G mobile phones 141
flat panel televisions *see* LCD (liquid crystal display) televisions; LED (light emitting diode) televisions
fluid design phases *see* experimental design phases
fluorescent lamps 53–5, *55*, **61**, 62
4G mobile phones 123
405-line televisions (VHF) 73, 74, 76–7, *85*
4K televisions 92, *92*, 93
front loading washing machines: dominant designs 25, 27, *28*, 29; environmental impacts 30, 31, 34; innovative designs 35–6, *36*
functions, unnecessary 167

Gaia Foundation 137
games consoles 77
gas lamps 47, *48*, 50
GEC televisions *85*
General Electric lamps 51, 52, 54, 55, 56, 58
GLS (general lighting service) incandescent lamps 51–2, *51*, 62, 63
Google 124
Google Glass headsets 141, *141*
green design 170–1
Greenfield, Susan 139
Grundig SVR video recorders 78
GSM (Global System for Mobile communications) 121, 122, 127
Guild, Lurelle 103

Haier washing machines 39–40
halogen lamps *see* tungsten halogen lamps
Hammer, Edward 56
hand-held vacuum cleaners 107, *107*
hazardous substances 90, 133–4
Hewitt, Peter Cooper 53
high definition television (HDTV): digital products 80, 81, 82, 83, 89, *89*; early systems 72, 79, 166
Hillman, William 5

home cinema kits 80
Hoover vacuum cleaners: bagless 112; tank *105*, 106, *106*; upright 100, *102*, 103, 104, *106*
Hoover washing machines: front loading 27, *28*, 32–3, *33*, 38; top loading 25, 26
Hoover, William 100
'horseless carriage' approach to design 24
Hotpoint washing machines *28*
housework 21, 22, 38, 39, 97, 115
Hurley Machine Company washing machines 22–4, *24*

IBM 58, 127
IDTVs (integrated digital televisions) 80, 82–3
Illinois Bell 121
incandescent lamps 48–53; carbon filament 48–50, *49*; dominant designs 52–3; efficiency **61**; life cycle assessments 62, *62*; life-time and cost **61**; tungsten filaments 51–2, *51*, 62, 63; tungsten halogen 52, *53*, 59, **61**, *62*, 63
inclusive design 177
incremental improvements 15
India 22, 27, 39, 97, 119, *120*
industrial design 15, 162–3; mobile phones 127, 129, 132, 163, 169; vacuum cleaners 101, 103–4, *103*, 108, 163; washing machines 32, *33*
innovative design phases 14, 16–17, 151–3; bicycles 9–10; lamps 53–5, 56–7, 58, 65–7, 153; television 79–81, 153; washing machines 35–6, 152–3
intelligent machines 35, 40–1
International Energy Agency 88
Internet connected products 40–1, 66, 82–3, *82*, 122, 123, 177
iPhone: design 129–30, *130*; environmental issues 135, 136; ethical issues 136–7; smartphones 123, 124, 125, 158
iPlayer 82
iRobot vacuum cleaners 112–13
ITV (Independent Television) 76, 77, 79
Ive, Jonathan 129

Japan Broadcasting Corporation 81
Jobs, Steve 129
Johnson, Alan 38
JVC VHS video recorders 78, *78*

Keep and Share knitware 44
Keracolor televisions 84

lamps 47–68; compact fluorescent (CFL) 56–7, *57*, 59, *60*, **61**, 63, *63*, 168; designs 52–3, 59; electricity use 60, 64, 65; environmental impacts and regulation 60–3, **61**, *63*; fluorescent 53–5, *55*, **61**, 62; future developments 65–7; incandescent 48–53, *49*, *51*, *53*, **61**, *62*, *63*; innovations 68, 153, 154, 155; LEDs 58, 59, *59*, *60*, **61**, 63, *63*, 158; prices 52, 56–7, 58, 61; social construction of technology (SCOT) 176; social influences and impacts 64–5; *Which?* tests 52, 55, 56–7, 58, 60–1, 66, 160

LaserDisc digital playback system 80, 159, *159*

laundry detergents 30, 31, 33, 36–7, 39

LCA (life cycle assessments) *see* life cycle assessments (LCA)

LCD (liquid crystal display) televisions 80, 81, 86, *87*, 88, 158

LeBlanc, Maurice 71

LED (light emitting diode) lamps: advantages and disadvantages **61**; energy labelling 63, *63*; life cycle assessments 62–3, *62*; technology 58, 59, *59*, *60*, 158

LED (light emitting diode) televisions 82, 86, *87*, 89, *89*, *92*, 158

Lee Maxwell Washing Machine Museum 24

legislation *see* Energy Labels; EU Directives

lessons for product developers 147–78; design options 157–65, 169–75, 176–8; failure avoidance 80, 165–8; knowledge and technologies 156–7; patterns of innovation 147–56; social influences and impacts 175–6; technology push and market pull 168–9

life cycle assessments (LCA): lamps 61–3, *62*; mobile phones 134–5; televisions 87; vacuum cleaners 114; washing machines 29–30, *30*, 32

life-cycle design 171–2

Loewy, Raymond 162

Lott, Tim 139

Magic Appliances vacuum cleaner *102*

manual washing machines 22, *23*

manufacturing processes 32, 51–2

Marconi Company 73

Marconiphone televisions 83, 88

market segmentation 15, 29, 83, 150

Matrix, The (film) 127

Maytag washing machines 25, *25*

McGaffey, Ives 98

mercury-vapour discharge lamps 53

metal halide lamps 55, 56

Michaux, Ernest 5

Michaux, Pierre 5

Microsoft 124

Miele vacuum cleaners *113*

minimum energy efficiency standards 63, 89, 114–15

mobile phones 119–42; design 124–32, *125*, *128*, *129*, *130*, *131*; disruptive innovations 169; environmental impacts 132–6, 171–2; ethical and political issues 136–7; experimental design 119–21; future developments 140–1; industrial design 163; operating systems 124, 177; pattern of innovation 141–2, 154, 155–6; prices 121, 126, 127, 131; social influences and impacts 137–40; sustainable design 172–3; technology 119–23, 141, 158, 166; *Which?* tests 122, 123, 124, 127, 160–1

Monotub Titan washing machines 35–6

Moore, Daniel 53

Motorola: 4500x car phone 125, *125*; clamshell phones 127, *128*, 129, *129*; DynaTAC 8000X mobile phone 121, 125–6, *126*, 169

Moulton, Alex 9, *10*

National Union of Students 67

neon lights 54

Netherlands 42, 80

Nilfisk vacuum cleaners 100

Nipkow scanners 72

Nokia: budget phones 131, *131*; early phones 126, 127–8, *128*, *129*; eco-profiles of handsets 134–5; end-of-life reuse 133; future developments 140; Microsoft takeover 124; smartphones 123

O2 mobile phone rating scheme 135–6

oil lamps 47, *48*

OLEDs (organic light emitting diodes) 65–6, 92, *93*, 158

1G mobile phones 121

1-watt initiative 88

online television services 82, *82*, 83

operating systems, mobile phone 124, 177

Panasonic televisions 81, *87*

patterns of innovation 44, 68, 93, 117, 141–2, 147–56

personal televisions 79, 166
personal video recorders (PVRs) 82, *82*, 89
phablets 130
Philips: compact fluorescent lamps 56, *57*; Green Focal Areas 172; innovative lighting 66, 67, *67*; LaserDisc system 80, 159, *159*; video cassette recorders (VCRs) 78
Phonebloks modular mobile phone 140
plasma televisions 80, 81, 86, 88
premature innovation 165–6
PremiumLight project 60
prices: affordable 161–2; lamps 52, 56–7, 58, 61; mobile phones 121, 126, 127, 131; televisions 74, 76, 77, 81, 83, 86, 161–2; vacuum cleaners 100, 103, 106, 109, 110; washing machines 26, 35
product differentiation 15
product placement 126, 127, 141
production costs 150, 152, 163, 167–8
PVRs (personal video recorders) 82, *82*, 89
Pye televisions 83

radical innovations *see* innovative design phases
RCA televisions 73, 75, *75*, 76
RCA-Victor televisions 83, *84*
RCA-Whirlpool vacuum cleaners 112
rebound effect 65, 174–5
recycling 90, 119, 133, 137, 173, 174
Regina stick cleaners 104
relative advantage 159–61
relevant social groups (RSGs) 13–14, 54–5, 175
reliability 164; televisions 77, 93, 164; vacuum cleaners 110, 115, 160; washing machines 28, 29, 152, 160; *Which?* magazine surveys 3
remote controls, television 77
rental of products 41, 43, 77, 78, 91, 178
replacement cycles 133, 176
resolution of televisions 81, 83, 153, 155
Rexair centrifugal vacuum cleaners 108
Ring, D.H. 120
Risler, Jacques 54
robot vacuum cleaners 112–13, 116, *116*
RoHS (Restriction of Hazardous Substances) Directive 90, 133–4
Rolls Razor washing machines 26
RSGs (relevant social groups) 13–14, 54–5, 175

S-Curve theory of innovation 151–2, *152*
Samsung: mobile phones 124, 130, *131*;

smart watches 141; televisions 87, *89*, 90, *92*; washing machines 35, *36*, 41
Sanyo washing machines 35, 40
satellite television 79, 80, 81
Sauerbronn, Karl von 5
Schaffer, Jacob 22, *23*
SCOT theory (social construction of technology) 13–14, 54–5, 175–6
Scott, Ray *26*
screen sizes 86, 140
set-top boxes 80, 82, 88
Sharp, Archibald 6
625-line televisions (UHF) 76, 79
Sky television 79, *82*
smart products: lighting 66; televisions 82–3, *87*, *89*, *92*; washing machines 40–1; watches 140–1
smartphones: definition 119; development 122, 123, 127, 140; environmental impacts 136; social impacts 138–9; *see also* Apple, iPhone
social construction of technology (SCOT) 13–14, 54–5, 175–6
social influences and impacts 175–6; bicycles 13–14; lamps 64–5, 176; mobile phones 137–40; television 90–1, 176; vacuum cleaners 115–16, 175; washing machines 37–9, 175
solar lighting 47, 59
Sony: Betamax video format 78, *78*; sustainability policy 90; televisions 77, *77*, 81, 84, *86*, 159
Sony Ericsson mobile phones 132, *133*
sound quality of televisions 81
Spangler, James 100, *101*
Sprengel, Hermann 50
standby energy use 88, 89, 135, 170
Starck, Philippe 12
Starley, James 5
Starley, John 6
stick cleaners 104, 106
successful product design 157–65
sustainable design 16, 172–3, 178
Sustainable Household (SusHouse) project 42–3
sustainable innovation 173–4
sustainable services 41–4, *43*, 67
Swan, Sir Joseph 49–50
system compatibility 164–5

take-back schemes for mobile phones 133
tank vacuum cleaners 100, *103*, 104, *105*, 106, *106*; *see also* cyclonic vacuum cleaners

Index

technological change theories 14–17
technological paradigms 15
teletext services 79
television 71–93; all-electronic systems *see* CRTs (cathode ray tube) televisions; analogue 76–7; design 83–7, 163, 170; digital 79–81, 82–3, 85, 158; electro-mechanical systems 72–3, *73*, 83; environmental impacts and regulation 87–90, *89*; future developments 92–3; innovations 79, 81–3, 91, 93, 153, 154, 155; prices 74, 76, 77, 81, 83, 86, 161–2; reliability 164; social influences and impacts 90–1, 176; *Which?* tests 2–3, 76–7, 79, 82–3, 93, 160
television standards 73, 76, 80, 81, 92
television studies 91
terrestrial digital broadcasting 80
Texas Instruments LEDs 58
text messaging 122, 139–40
3D television 79, 82, 167
3G mobile phones 122–3
top loading washing machines 25, 27–8, *28*, 30, 31; *see also* twin-tub washing machines
Total Recall (film) 141
touchscreen technology: Apple iPhone 123, 129–30, *130*, *131*; dominant designs 131, *131*, 132, *132*; IBM Simon 127; operating systems 124; Samsung Galaxy 130
tumble driers 37
tungsten filament lamps 51–2, *51*, 62, 63
tungsten halogen lamps: advantages and disadvantages **61**; energy labelling *63*; life cycle assessments 62, *62*; technology 52, *53*, 59, 153; withdrawal from the market 63
Twigger Holroyd, Amy 44
twin-tub washing machines 26–7, *27*, 31
2G mobile phones 121–2

UHF 625-line televisions 76, 79
ultra high definition televisions 92, *92*
United States of America: television 73, 74, 75; washing machines 25, *25*, *26*, 27, 34, 38
unnecessary features and functions 167
upright vacuum cleaners: dominant designs 103, 104, *105*, *106*; early designs 100, *101*, *102*; prices 100, 103, 109, 110, 161; *Which?* tests 106; *see also* cyclonic vacuum cleaners
urban electric lighting 50

user-centred design 177

vacuum cleaners 97–117; cyclonic 108–11, *109*, *110*, *111*, *113*, 168; disruptive innovations 107, 112, 158; dominant designs 100–7, *103*, *105*, *106*, *107*; early designs 98–100, *99*, *101*, *102*; environmental impacts 114–15; future developments 116; industrial design 163; innovations 112–14, 116, *116*, 117, 154, 155; prices 100, 103, 106, 109, 110, 161; social influences and impacts 115–16, 175; *Which?* tests 104, 106, 109–10, 112–13, 114, 160
VCRs (video cassette recorders) 77, 78–9, 80, 82
VHF 405-line televisions 73, 74, 76–7, *85*
video cameras 77
video 'format war' 78–9, *78*
video projectors 86
Vodafone 121, 123, 136

Wall Street (film) 126
WAP (wireless application protocol) 122, 123, 127
washing machines 21–44; detergents 30, 31, 36–7, 39; dominant designs 25–9, *25*, *26*, *27*, *28*, *29*, 150; environmental impacts 29–30, *30*, 37, 38–9; environmental labelling 31–4, *33*, *34*, 171; experimental designs 22–4, *24*, 148, 154, 155; industrial design 162–3; innovations 35–6, *36*, 39–41, *40*, 152–3; ownership 22, 25, 41; prices 26, 35; social influences and impacts 37–9, 175; sustainable laundry services 41–4, *43*; system compatibility 164–5; *Which?* tests 2, 25–6, 27, 31, 33, 34, 35–6, 160
water-free washing machines 40, *40*
wearable technology 140–1
WEEE (Waste Electrical and Electronic Equipment) Directive 90, 133
wet and dry tank cleaners 107
Which? magazine 1; analogue television 76–7; bicycles 9, 12; cordless vacuum cleaners 114; digital television tests 80–1, 82–3, 86, 92, 93; energy use, televisions 88–9; evaluation methods 2–3, 159, 160–1; Internet connected lighting 66; lamps 52, 55, 56–7, 58, 60–1; laundry detergents 36–7; mobile phones 122, 123, 124, 127; smart watches 141; vacuum cleaners 104, 106, 109–10, 112–13; video cassette

Index **191**

recorders (VCRs) 78, 79; washing machines 25–6, 27, 31, 33, 34, 35–6, 160
widescreen television 79, 85, *86*
Windows Phone 124
women's work 21, 22, 38, 39, 97, 115

Xeros washing machine 40, *40*

Young, W.R. 120

Zanussi washing machines *29*, 32
0G mobile phones 120–1
zero waste 137
Zworykin, Vladimir 73

427545